W. Douglas Smith is an environmental scientist with thirty years of field experience with the U.S. Environmental Protection Agency (EPA). In addition to conducting hundreds of investigations, he wrote the most widely used EPA training manual and trained new inspectors and investigators for EPA and the National Enforcement Training Institute (NETI) in Denver, Colorado. He led multi-media teams that monitored regulatory compliance and "environmental management systems" under domestic and international operating standards. From 1996 to 2005, Mr. Smith served as the EPA liaison to the United Nations and the World Bank Institute. Together, they assisted nations in developing their own environmental protection programs. Mr. Smith also served on the Board of Directors for the Seattle chapter of the United Nations Foundation. Additionally, he owned an international adventure travel company for more than thirty years, which allowed him to explore numerous ancient civilizations, cultures, and remote regions around the world. His intimate experiences with environmental laws, multi-national corporation management systems, and the development of environmental programs provide him with a unique perspective on sustainability, human behavior and the multiple existential crises threatening today's global civilization.

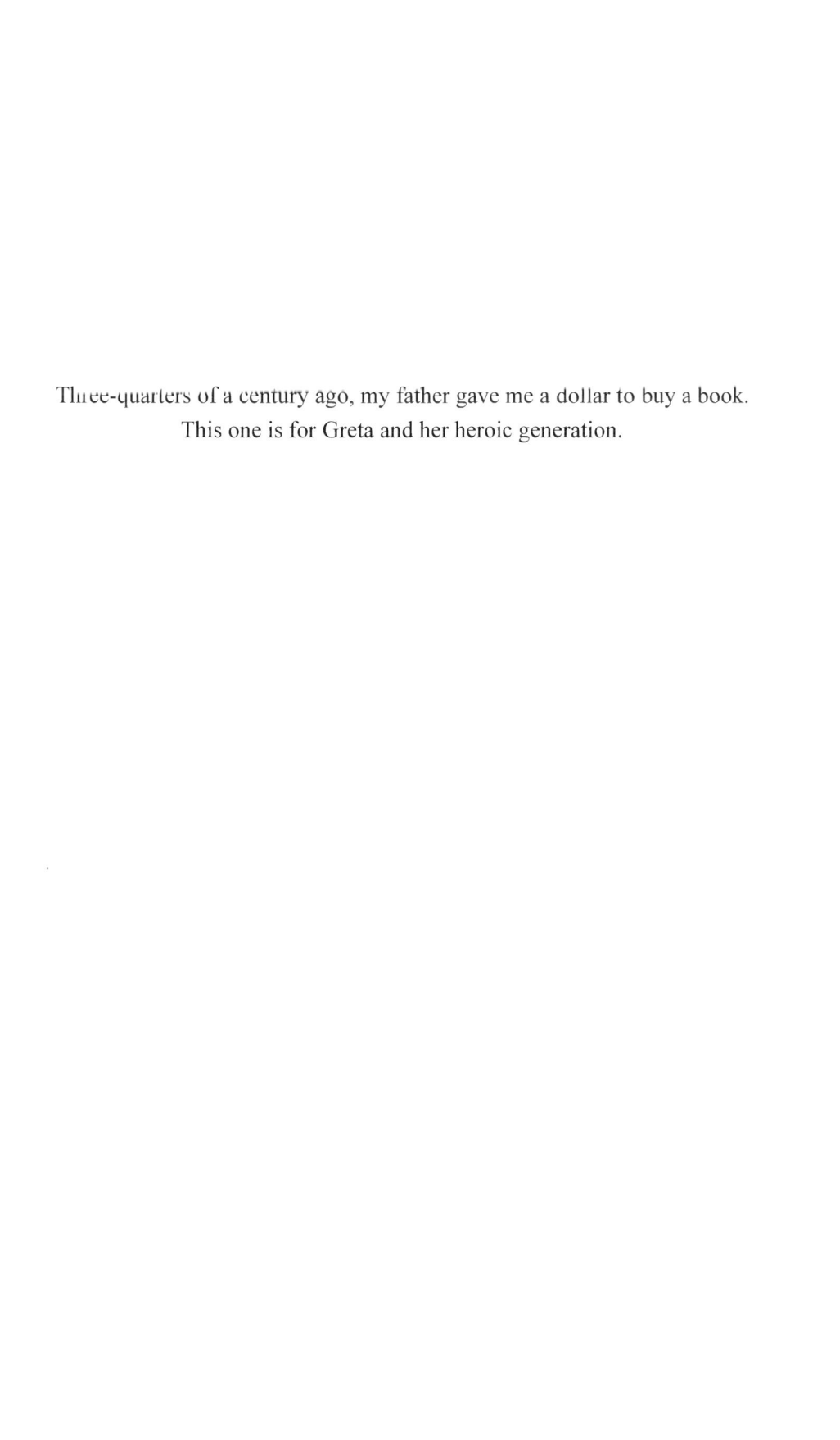

Three-quarters of a century ago, my father gave me a dollar to buy a book.
This one is for Greta and her heroic generation.

W. Douglas Smith

FUTURECIDE

AUSTIN MACAULEY PUBLISHERS
LONDON * CAMBRIDGE * NEW YORK * SHARJAH

Copyright © W. Douglas Smith 2024

All rights reserved. No part of this publication may be reproduced, distributed, or transmitted in any form or by any means, including photocopying, recording, or other electronic or mechanical methods, without the prior written permission of the publisher, except in the case of brief quotations embodied in critical reviews and certain other non-commercial uses permitted by copyright law. For permission requests, write to the publisher.

Any person who commits any unauthorized act in relation to this publication may be liable to criminal prosecution and civil claims for damages.

The story, the experiences, and the words are the author's alone.

Ordering Information
Quantity sales: Special discounts are available on quantity purchases by corporations, associations, and others. For details, contact the publisher at the address below.

Publisher's Cataloging-in-Publication data
Smith, W. Douglas
Futurecide

ISBN 9798889100324 (Paperback)
ISBN 9798889100331 (Hardback)
ISBN 9798889100355 (ePub e-book)
ISBN 9798889100348 (Audiobook)

Library of Congress Control Number: 2023916143

www.austinmacauley.com/us

First Published 2024
Austin Macauley Publishers LLC
40 Wall Street, 33rd Floor, Suite 3302
New York, NY 10005
USA

mail-usa@austinmacauley.com
+1 (646) 5125767

Table of Contents

Introduction   10

Futurecide   15

   Chapter 1: The End of the Holocene   17

   Chapter 2: Lessons Learned and Lessons Forgotten   34

   Chapter 3: Connecting the Dots   59

   Chapter 4: Sociopathic Genes and Politics   67

   Chapter 5: AI and Manipulation of the Truth   85

   Chapter 6: The Last Roundup   93

   Chapter 7: System Thinking   111

   Chapter 8: The Decadence Gene   122

   Chapter 9: Bugs on the Windshield   130

   Chapter 10: The Paradox of Two Realities   136

   Chapter 11: Government vs Governance   156

   Chapter 12: Deregulation Gives Corruption License   171

   Chapter 13: Overpopulation   181

   Chapter 14: Gross Domestic Deception   189

   Chapter 15: Precautionary Principle and Security   207

   Chapter 16: Greed, Sex, and Equity   219

   Chapter 17: Equity   225

   Chapter 18: Stewardship   236

Chapter 19: Women and Civilization   *239*

Chapter 20: Discounting the Future   *247*

Retrospective   *258*

Bibliography   *260*

Praises for *Futurecide*

Doug Smith intertwines his personal stories as an EPA Senior Compliance Investigator with updated research to give the reader a clear picture of our human predicament. Readers will be left in awe of the planet and concern for our future.
– Dr. Sibylle Frey, BSc MSc, Director of Production, Millennium Alliance for Humanity and the Biosphere.

The synopsis is great as is the book…a big contribution.
– Peter Carter (Senior IPCC Report reviewer)

My first impressions were born out. This is well written, thoroughly researched, and the most comprehensive approach I've seen on the topic. I enjoyed all the personal touches and accounts of your own wide and valuable experience in this area.
– Nicholas Gier, Emeritus Professor of Philosophy (35 years), University of Idaho, Moscow, Idaho.

Introduction

"Everything has changed except our way of thinking." (Albert Einstein).

The future is threatened in a way that is unprecedented in the history of mankind. This is not a book of tedious charts and graphs. It is a narrative of my investigation of four things: the reality and scope of climate change, the consequences of a collapsing environment, how human behavior determines how we react, and the indictment of those who perpetrated the growth and continuation of the crisis. It is not chronological because investigations are a reiterative process of cross checking, going back for another look and making connections that lead to new discoveries. The reader accompanies me on a personal journey of discovery. What I learned shocked me as much as it probably will you. Our species has responsibilities that cannot be ignored.

We forget that the environment provides 100% of the goods and services that allow civilization to exist. Even before the 1960s it became clear that Anglo-American economic policy was failing to sustain the environment within planetary limits. The business psychology of a consumer economy discounts the social and environmental pillars necessary for a sustainable civilization. An experience I had in Indonesia may help to illustrate what I mean:

"It's just business," the man said.

The man was the Chief Financial Officer (CFO) of a British Petroleum (BP) petrochemical plant on the island of Java. I was in Indonesia to train new inspectors for the Indonesian government. The plant had only been completed a few years earlier. It produced plastic beads that other companies used for injection molding everything from car bumpers to toys. From the time they commenced production, BP had not once operated their wastewater treatment plant. Waste disposal wasn't a topic senior management thought about. Increasing production and profit was their sole focus. The result was to let their waste accumulate and rot in the hot tropical sun or to pump it directly into the South China Sea.

The plant warehouse was larger than several commercial airplane hangars. Piled against the outside of the building were many thousands of bulging drums, containers of chemically reactive spent solvents, explosive peroxided ethers, flammable oils, and tens of thousands of bags and containers of toxic waste. The soil was a stinking, rainbow-colored ooze from rusted and leaking containers. The waste wasn't only a regulatory issue, it had the imminent potential to blow up the entire plant.

A few hundred meters away were the shacks where the plant workers and their families lived. The unsecured waste site was their children's playground. I wanted to know why toxic, flammable, and explosive waste had not been properly disposed.

The CFO said our visit was the first time the plant had been inspected, as if that was the only motivation to manage their waste. BP plant management promised they would rectify the problem immediately.

Indonesia didn't have a designated hazardous waste site at the time. Despite that fact, the next day I was notified that the waste was gone. The BP representative told me that my 'envelope' would be waiting in the plant office. I declined the offer of an 'envelope' (cash) and advised my student inspectors never to entertain even the suggestion of a bribe or the entire purpose of their jobs would mean nothing.

Where did those tons of hazardous waste go? As an environmental diplomat I had to leave the follow-up and the 'envelope' issue to official Indonesian government policy.

A few days later the acting Director of Enforcement for the new Indonesian Environmental Impact Management Agency addressed the graduating class of new inspectors. In full uniform with braid and regalia, he announced that graduating inspectors were now certified government agents. He continued that instead of companies offering inspectors 'envelopes' or gifts like golf clubs, they could now demand memberships to golf resorts. The reader might find this shocking, but it was far from the first time I experienced corruption, bribes, or threats.

Corporate culture is very different from the biological human community. There is no question that civilization needs a strong and innovative business community, but some businesses view protecting public health and the environment as an unnecessary burden. That attitude is often proportional to the size and nature of the corporation. The for-profit corporation's entire *raison d'etre* is to make a profit. That is specifically codified in their corporate charter. In this strange way, the corporate charter encourages sociopathic behavior. The priorities of society and the environment tend to be viewed as a threat to their bottom line. There is little motivation to consider the wellbeing and security of the public on a par with making a profit unless doing so improves their quarterly report. Despite this, a US corporation is granted the same legal rights and protections as the biological human community. The result is inequity in political representation via 'envelopes' addressed to political campaign chests. This difference in perception often conflicts with social and environmental priorities.

The basic chemistry and physics necessary to understand global warming was established nearly two hundred years ago. The rules that maintain planetary systems within a habitable environment are fixed by the laws of thermodynamics and bio-geochemical systems, not profit margins. When civilization ignores these facts, it is guaranteeing catastrophe. The BP example illustrates how the myopic pursuit of greater profits blinds a business to their wider responsibilities. Unconstrained economic policies are now destroying the environment and ignoring the fact that civilization cannot exist without it.

Earth is a closed system. The consumption of resources is now more than the Earth can sustain. It is vital to find a balance between the economy and social wellbeing within the sustainable limits of planetary boundaries. There is no alternative. The global economy must adapt to prevent environmental collapse. It will demand a massive and rapid cultural transition. But it will not take place without unprecedented multi-national government coordination, support, and leadership.

What are my qualifications for writing this book? I grew up on the edge of the Oregon wilderness to become a teacher, an environmental scientist, an explorer, and a senior federal investigator for the US Environmental Protection Agency (EPA). After college, I taught science for the Bureau of Indian Affairs on the Navajo reservation. There I learned that some civilizations, far older than ours, strived to live within the boundaries of nature. The shiny objects of today's global civilization corrupt those connections, tending to make us believe we are exceptional and apart from the natural world.

Ethnic prejudices and the abuses inflicted on Native Americans angered and confused me, so I pulled up stakes, moved back to the northwest, and was hired by the US EPA as an unleaded fuels inspector. Within a few years, I became EPA's first Senior Compliance Investigator, authorized to conduct compliance inspections and investigations for all laws under EPA's mandate.

I conducted investigations of small and large businesses. I put teams together to monitor federal facilities, and some of the world's most powerful corporations. Part of those duties included reviewing business and corporate management practices. Later, I was assigned to train new agents. I wrote an environmental compliance manual that was used throughout the EPA and the National Enforcement Training Institute (NETI) in Denver. By the ninth edition, it was also used in dozens of other nations.

In addition to my regular EPA duties, I became an investigator for EPA's Suspension and Debarment Program and worked with an attorney who could not be intimidated by the high paid lawyers that corporations pitched against the government. We conducted follow-up investigations of those previously found in violation of EPA civil or criminal statutes. Further violations meant they could be suspended from doing business with the US government (the world's largest consumer) for a specified period. If repeat violations were severe enough, a person or business could be debarred from doing business with the government for longer periods, including in perpetuity.

During the last years of my three decades with the EPA, I was assigned to work with the United Nations and World Bank Institute. We helped other nations develop their own environmental programs.

In addition to my work with the EPA, I owned an adventure travel company that led expeditions to some of the more remote regions of the world. We promoted the concept of Eco-Trekking to help people understand humanity's connection with the natural world. Those experiences allowed me to intimately witness the patterns and evolution of our modern global civilization, and the diverse connections with the environment that sustains it.

The Laws of Ecology
(Ernest Callenbach)

All things are interconnected.

Everything goes somewhere.

Nothing is free.

Nature bats last.

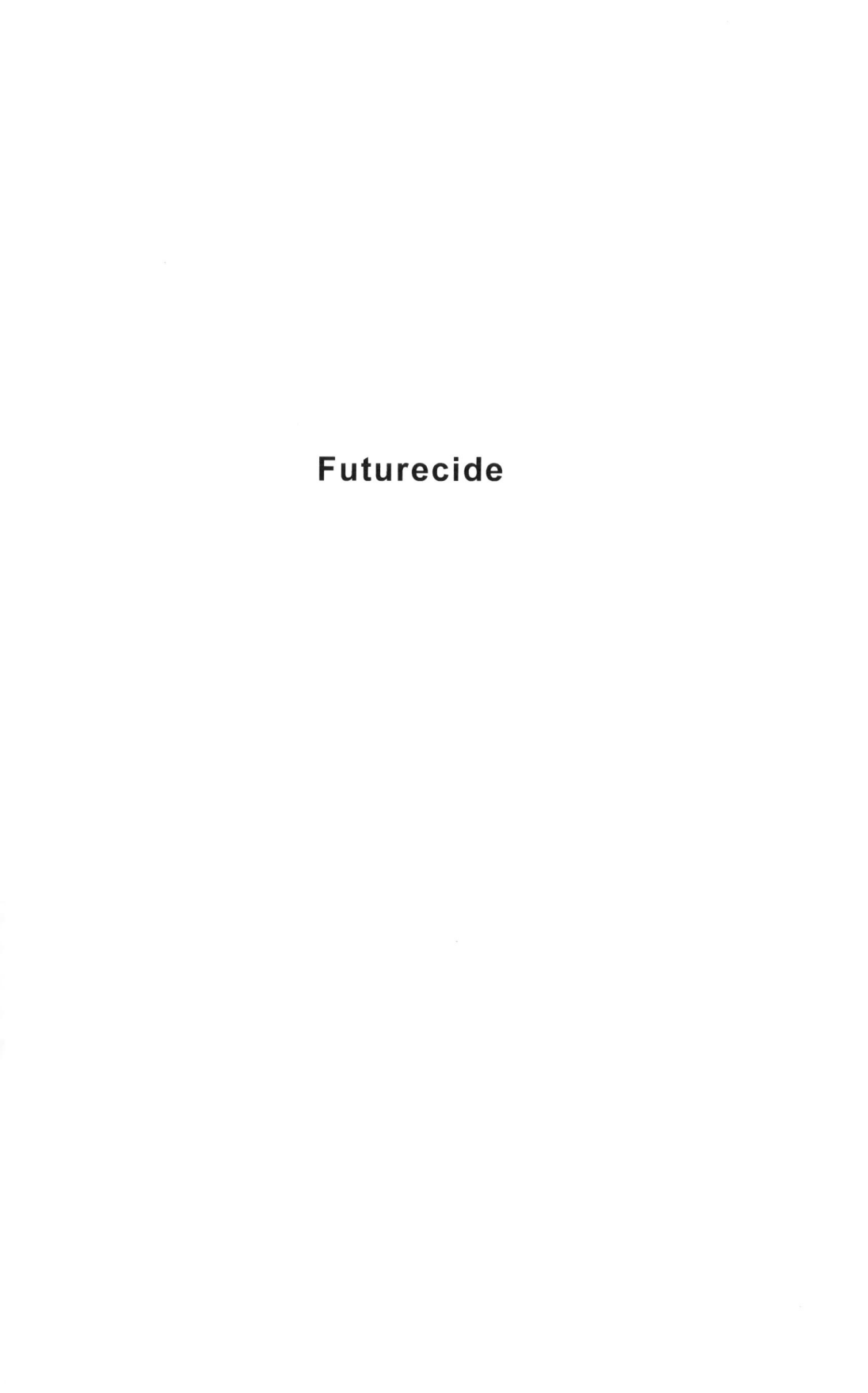

Futurecide

Chapter 1
The End of the Holocene

Futurecide is the knowing and willful setting into motion the destruction of planetary bio-geochemical systems necessary to sustain future generations of life on Earth. What is happening today is not genocide or ecocide. It is global and progressively irreversible suicide.

The past twelve thousand years of relative stability has passed. From this moment forward, civilization will exist in constant crises and extreme environmental disruption. The perpetrators of futurecide have lied, cheated and defrauded humanity on a massive scale. A few powerful elitists continue to commit this ultimate crime against humanity and our earthly habitat. The scope and magnitude of their crime has been intentionally distorted and remains poorly communicated or understood.

The biosphere is undergoing a global mass extinction, amplified by the behavior of a handful of individuals and businesses. The linkages between global warming, climate change, soil loss, finite mineral shortages, a food and water crisis, disease, political and economic insecurity, conflict stressors, and massive refugee migration are becoming clearer. They all contribute to projections that eight in ten people alive today will die prematurely. Life expectancy in the United States has already declined by two years.

A competitive drive for sociopathic power has found a prominent niche in politics and serves as the driving force behind a deregulated economy. This has led to political and legal subservience to the corporate engines of capitalism, not human or environmental wellbeing. Unconstrained corporate greed and irrational consumption are no longer sustainable.

Tyrants and sociopathic leaders were not uncommon in history. Even the greatest civilizations encountered economic instability, injustice, corrupt politics, class struggles, resource shortages, and threats from inside and outside their borders. Every civilization in history depended upon the stability of the environment to provide food, water and the goods and services necessary to prosper. Yet every civilization in history failed. Was the glitch something built into civilization or in *Homo sapiens* or both?

Today humanity faces the absolute limits of planetary resources. It may also be approaching a point beyond the human capacity to adapt. If that is the case, *Homo sapiens* may have created a world it is not genetically programmed to thrive in.

These questions first came to mind some time ago when I spent a very cold night atop the great pyramid of Khufu on the outskirts of Cairo. I shivered alone, on one of the seven wonders of the man-made world, wondering what could have brought this mighty civilization to ruin. Egypt ruled their part of the globe for nearly three thousand years. I tried to imagine what it must have been like when they ruled an empire that stretched the length of the Nile from the Mediterranean to the heart of the African continent. They must have felt omnipotent or at least invulnerable. They looked up at the same Milky Way that shimmered above me like some deity had cast billions of diamonds onto a black velvet tapestry. Did they believe that tapestry was somehow put there just for them?

Over the decades I've traveled the globe and visited most of the great remnants of human history. But it was at that moment, half frozen, hugging my knees and shivering on a manmade sepulchre of ego and stone that it became clear. My species was not the center of the universe. We weren't invulnerable. It wasn't all about us. If we failed to stay in sync with what was going on around us, there were consequences. As I listened to my teeth chatter, I tried to occupy my mind with what I could remember about the rise and fall of the great civilizations in history.

Despite our hubris, on a geologic time scale, our species is an insignificant experiment, and not as divinely blessed as we like to imagine. Our survival is not guaranteed. Despite differences in technology, every civilization before us had to deal with the same social, economic, and environmental problems. Despite their similarities and the recorded experiences of history, not one civilization has survived in the twelve thousand years since the last Ice Age. Was there a pattern that led to collapse? Where was the fatal glitch?

In the 21st century, the answers to those questions become far more important. A host of modern scholars have studied how many of the great civilizations transitioned through nearly identical phases. Instead of the traditional linear approach to history, these modern scholars took a three-dimensional approach. They showed how the linkages between the social, environmental and economic aspects of civilization determined if they were sustainable or not. I poured over the works of Yuval Noah Harari, author of *Sapiens, and Homo Deus*; Jared Diamond and his trilogy, *The Third Chimpanzee, Guns, Germs, and Steel*, and *Collapse*; and William Ophuls' book *Immoderate Greatness—Ecology and the Politics of Scarcity*. I had bits of the puzzle, but the picture was incomplete. As with any puzzle it helps to establish the key pieces first.

What were those common phases? The first pioneers had to find an environment with a mild climate, rich soil, storable plant grains, and a dependable water supply. Without these basics it would be impossible to establish a permanent settlement. The pioneers were soon followed by the builders. The builders constructed permanent homes, gardens, granaries, public buildings and fortifications. After that came the schools, philosophy, literature, and the law. This phase was sometimes called the age of learning that fueled a renaissance of art and culture. The expansion of greater skill and knowledge was usually followed by a corresponding increase in power and wealth. Here there seemed to be a diversion from development and national focus to amusement and instant gratification.

As wealth increased, decadence, greed, and corruption often followed. This last transition seemed to be where most civilizations became the most vulnerable. Three forces began to compete for leadership. One group wanted to hold to tradition. Another group would seek to hold onto or increase their power. The third group usually consisted of the scholars and a business community that saw both threats and the opportunities of change. While these forces were competing for control, the general population was preoccupied with the complexities of everyday life.

The transition through these phases was not linear but a complicated web of linkages and events. The influences of leadership, internal and external dynamics and the environment all had their impact. Wealth and power might seem like security, but also fuel growing hubris and sense of entitlement. Generations following the builders and philosophers might not exercise the same patient foresight of their forefathers. As a multi-generational aristocracy developed, they would begin to believe they were superior, sometimes without actual merit. Unconstrained power and wealth tended to stratify society. Those born to privilege reaped greater benefits, while the underclass that performed the work, received an ever-diminishing portion.

At some point between wealth and decadence, there appeared to be a perilous juncture in the evolution of civilizations. Struggles for leadership and the desire to concentrate authority begin to dominate politics. This pattern is recorded in the histories of most of the great civilizations. As a society becomes more unstable, rational behavior begins to unravel. Decisions might not be based on general wellbeing or the accumulation of knowledge and wisdom. Unconstrained wealth and decadence creates an atmosphere of intrigue for the concentration of power and competition for authority. These struggles often threw the general population into chaos. Society begins to fracture and revert to a more primal, almost tribal state that further corrodes the ability to apply knowledge rationally.

The *Homo sapiens* species is fundamentally a highly social, hunter-gatherer. We are also territorial and an apex predator. Early human groups or bands were small and held together with tight knit bonds and familial ties. The stone-age mind created stories and myths that helped form the rules and norms that held those small societies together. Security was built upon mutual dependence and an intimate knowledge of others that could be relied on. That trust meant that the threats and obstacles of existence would be met as a team. Sociopathic or psychopathic members that proved problematical were shunned, punished or cast out (Boehm, *Moral Origins*, 2012).

Small group integrity, trust and interdependence benefited human survival for hundreds of millennia. Those same intimate, small group norms of behavior may now be maladaptive to the pace and complexity of eight billion people struggling to share a single habitat (Earth). Instead of striving for a stable society and wellbeing, a consumer economy and competition for power are driving today's global civilization off an extinction cliff.

The divisions of labor and specialization may be fracturing the intimate bonds that human society needs to build trust and maintain security. Instead of shunning the sociopath's anti-social, competitive drive for power, sociopathy has found a prominent niche' in politics and as the driving force behind deregulated Anglo-American capitalism. The norms of order, conformity and loyalty that held small bands of humanity together are becoming subservient to the insatiable engines of consumerism.

The 20th century Anglo-American economic model built an economy on greed and overreaching consumption. The same forces that divided Rome, Athens and Egypt can be seen in 21st century America. Alienation between the pursuit of wealth, tradition and a younger generation striving to adapt to change are now three groups approaching open warfare. Our global civilization now faces a human behavior crisis in addition to a bio-geochemical crisis.

If our 21st century, global civilization is to prevent its extinction, we must first define the problem. What we know is that an elite minority is controlling the agenda. They are knowing and willfully setting into motion the destruction of systems necessary to sustain future generations of life on Earth. They are literally destroying the future wellbeing of humanity. What else do we know and what do we need to know?

In the following chapters I'll address the three social forces competing for control. There are those seeking to gain and maintain wealth. There are those who want to return to the old stories and myths of Americana. And there are those seeking change to what they see as an interconnected mix of existential social, economic and environmental crises. The perpetrators of futurecide have lied, cheated, and defrauded humanity on a massive scale. Civilization's footprint threatens the security of all life on Earth. Economic and politically driven fabrications continue to drive disruptions in the global environment that are best described as suicidal.

The rise and fall of civilizations serve to illustrate the paradox we face. We are genetically a hybrid Cro-Magnon, adapted to living in small familial bands trying to survive in an overcrowded, rapidly evolving global civilization. We used to be a few people living on a big planet. Today we are too many people living on a small planet. To survive we must change our behavior, our lifestyles, and the economy to a more rational and sustainable existence. We must do this because the laws of nature are not forgiving, and the clock has run out. We can no longer save all of humanity but what we do and how quickly we do it will determine how much of humanity we can save.

The undimmed eyes of the younger generation see democracy threatened by a growing populism, economic greed, overwhelming technology and an environment collapsing before their eyes. Their rights and freedoms are diminishing. Their voices are stifled as democratic principles are threatened by authoritarianism. More than a billion people are slowly starving. Refugee numbers increase daily as crops fail, storms rage, and water shortages spread over entire continents. Two billion people are without jobs that pay enough to buy a home or have hopes of ever owning one.

On the 23rd of September 2019, a diminutive sixteen-year-old girl from Sweden addressed the United Nations Climate Action Summit in New York City. Greta Thunberg was already a hero to young people around the world for her courage in speaking truth to power. Her warning was stark. "The eyes of all future generations are upon you. And if you choose to fail us, I say: we will never forgive you."

For a moment, she paused. Her lips tightened as she glared at the audience with tears of outrage. "You have stolen my dreams and my childhood with your empty words and yet I'm one of the lucky ones. People are suffering. People are dying. Entire ecosystems are collapsing. We are in the beginning of a mass extinction and all you can talk about is money and fairytales of eternal economic growth." Her words are now part of history, not because they were spoken by a child, but because they were true.

The younger generation has tried to tell their parents what they learned in school. Some parents respond with ridicule or worse. "You can't tell me a couple degrees of warming is such a big deal," or "Even if the environment is in trouble, it's happening so slowly it won't get serious for decades." The cynics give cheeky grins and proclaim, "I'll be dead by then." Some flippantly pass responsibility off to politicians and then don't vote.

A third of America found Greta's message difficult to believe, though the dire changes she spoke about surround us. To rich countries the world seemed to be ticking along. The view out the window looked pretty much like it did a year ago, and the year before that. The rich didn't feel the pangs of hunger and privation. They were confident they could shelter through any storm—it was nothing they couldn't handle.

Winston Churchill once said, "America always does the right thing after it has tried everything else." But there isn't time for everything else. Humanity must get it right in one shot. Half measures will not prevent catastrophe. In truth, half measures only amplify burgeoning disaster.

There is a hierarchy to suffering in this crisis. The poor, children, women and elderly are the most vulnerable and suffer first. Powerful elites are the most responsible for today's multiple, interconnected crises, but they don't feel or see it as a crisis. They don't see how droughts affect electricity or that high energy costs may mean a freezing bedroom in winter or the inability to buy gasoline to get to work. They don't see how a single health issue can drive an entire family into poverty. They don't feel rising food prices. They don't feel bigotry and injustice. They don't see how the lack of an education makes people blind to opportunities the wealthy take for granted. They don't see how the lack of childcare or an unplanned pregnancy handicap that child and parent for the rest of their lives.

The days, months, and years of quiet public desperation go unheeded. Wealth insulates the rich from all that. Humanity lives in a world divided between the 1% that have everything and the 99% struggling just have something. People become locked into divergent perceptions of reality, unable to see the urgent need for action. Some may cynically believe meaningful action is impossible.

Every civilization in history encountered the same problems. Every civilization depended upon the stability of the environment to provide food, water, and the goods and services to maintain their economy. Failing to recognize the social, environmental and economic connections to a sustainable future was and is a sure path to collapse.

Alexander thought he had conquered the world, so did Genghis Khan, the Romans, and the British Empire. Vast as their empires were, they were not truly global. Like pulling your finger out of the water, the void they left was quickly filled and life went on. In the past, when the local environment was damaged or depleted, there were always new frontiers to explore and develop, new resources to tap. The world used to be rich, vast and mostly unoccupied. In the 21st century, there are no more frontiers—civilization has taken up all the slack in the environment.

Frank Unger may have said it best in his 2004 book *Friends in High Places*. "It's hard to remember to drain the swamp when you're up to your ass in alligators." The 'alligators' distract us from what has become the greatest threat magnifier humanity has ever faced. The swamp is a rapidly warming planet, overpopulation and overconsumption. Civilization's footprint is responsible for an unprecedented loss of species diversity. That loss of diversity compromises the planet's resilience to human insult and started the most rapid mass extinction in Earth's history.

Peter Turchin, Professor of Ecology and Mathematics at the University of Connecticut, wrote that the very success of a civilization was tied to its downfall with mathematical certainty. He reviewed the past ten thousand years of human history and found that there were several law-like propositions of general ecology that appeared applicable to the story of humanity. Those patterns of human nature and the history of civilizations have repeated with almost mathematical predictability. If we understood these patterns, it might help understand the progress or decline of any species, including *Homo sapiens* in the past, present, or in the future.

Turchin's research found that competition for lofty status and subsequent decadence tended to corrupt the priorities of leadership and the social stability necessary to maintain a smoothly functioning society. The pampered and distorted values of the highly privileged, detract from the fundamental housekeeping necessary to sustain an orderly and just society. In the great civilizations of the past, leaders became lords and lords became Caesars who dreamed of becoming gods. Without checks and balances the authority of leadership tended to concentrate to a few, or even a single individual. The future of that civilization would then become vulnerable to the whims of the privileged and their distorted perceptions of reality. This could not be more evident today.

A little research found this to be a form of metacognition called the Dunning-Kruger effect. This was first used to describe people with low abilities who did not possess the ability to recognize their own incompetence. Psychologists soon learned that people with normal, or even above normal intelligence could also believe they knew more than they did. Wealth and privilege shelter a person or group of people from the normal consequences of their behavior. Because of this lack of feedback, excessive privilege in the most elite may form a delusional bias, making them incapable of recognizing their own incompetence. They might believe their accomplishments and perceptions were better than ordinary people. They might believe they would be successful at tasks they didn't have the skills or knowledge to succeed at. We know this happens with individuals. I wondered if it might also be true of nations.

It seems likely that an advanced civilization might also become sheltered from reality and lose touch with real-world consequences. Throughout history, leaders and self-aggrandizing elitists often set personal priorities above the wellbeing of the general population. Societies evolved into classes of privilege, and those cut off from it. If the elite had never missed a meal, it would be hard for them to understand those who have never been full.

There is little argument that a wise and benevolent leader can bring about enormous progress, but it can take only one tyrannical, decadent, or corrupt leader to bring it down. In Rome, the Senate provided a balance of power but as power concentrated to a single leader, it would only take a Caligula or Nero to bring greatness to chaos.

Economist Jeffrey Sacks warned in *The Age of Sustainable Development* that our global civilization is now facing the juncture between hedonism and sustainability. The stressors of injustice, socio/economic stratification, and arrogance are ignoring the three pillars that sustain civilization. The cleric Thomas Malthus said collapse was the result of war, pestilence, and famine, but what are the drivers behind these three apocalyptic forces? The patterns described by Turchin, Diamond, Sacks, Harari, Malthus and others were too strong to ignore.

Today, I'm a great-grandfather. My children ask me if we really are in danger of collapse. Why was my generation leaving a world in chaos, with depleted resources, pollution, and storms not seen in millions of years? They deserved an honest answer. My investigation to answer their question turned out to be the most far reaching and important of my career.

Now more than sixty years after my freezing desert night atop the great pyramid of Khufu, our burgeoning global civilization appears to be repeating precisely the same familiar patterns that drove all previous civilizations to collapse. This global civilization arrogantly teeters on a knife-edge. Have wealthy nations become elitists, nurtured in a bubble of privilege and sheltered from the consequences of poor choices? Will humanity choose decadence and a dystopian future or begin a new renaissance of social wellbeing and a sustainable environment? How important is it, at this precise moment, to choose the right path? Is there enough resilience in society, the economy, or the environment to recover if we choose the wrong one?

What if we submit to the same neolithic compulsions that led to the failure of every great civilization before us? How does a civilization guard itself from the delusional sociopath who focuses on power and privilege instead of the sustainable wellbeing of everyone? What is it that motivates so many to obediently follow the demigod? Why do we crucify the Samaritan and glorify the autocrat? Power and celebrity can distract the masses from the faults of leadership. The buildup of Hitler's Germany before World War II proved that. Is this the calculus in the human genome Turchen described? Does some innate residue of the human genome conflict with building and maintaining a sustainable civilization? How does a civilization guard itself from itself?

The Roman Philosopher Seneca said, "Luck is what happens when preparation meets opportunity." For nearly a million years, our species barely survived a tempestuous environment. For at least three thousand centuries, in wave after wave, *Homo sapiens* explored the earth, seeking the most hospitable habitats to settle. Our progenitors beat back predators, hunger, and hostile environments. We befriended the wolf and grass eaters. The environment was a harsh master, but it made us adaptable, resilient, and smarter.

Through the eons, we parried and thrust our way until we finally gained dominance. Those primal instincts and sense of community helped us survive in an untamed wilderness. Those same predatory, small group, communal instincts are now unsuitable for a global civilization of eight billion. In the end, only our species remains to be tamed.

At Delphi, there is the inscription 'know thyself'. I wondered if the evolution and decisions families make was like the evolution of past civilizations. I began to reflect on the similarities between the rise of civilizations and the history of my own family.

In 1995, my father and I visited my grandmother in Oklahoma, to celebrate her centennial. We sat around the dining room table for hours, listening to stories about her life and experiences. She spoke about homesteading and carving a living out of the Rocky Mountain west. She told stories about her childhood. She remembered listening to her best friend's parents talk about the Civil War, and how her friend's great-grandmother told her how she cheered the Lewis and Clark Corps of Discovery when they returned to St. Louis. My grandmother reminisced about gold mining, life in the camps, finding a place to settle, raising my father, the influenza epidemic of 1918–1920, and how the irrational greed of booming prosperity led to the Great Depression.

She spoke about how efficient so-called 'civilized nations' were at killing each other. She reminisced about building a home and sacrificing so that my father and his sisters could get an education. She told us how Roosevelt's 'New Deal' pulled America out of the depression. She spoke about how electricity, the telephone, radio, television, cars, airplanes, the atom bomb, and cell phone had each completely changed the world. My grandmother's story comprised almost half of the history of America. Her story paralleled the pioneers, builders, and learning phases of so many civilizations. Did mine?

My father was part of a generation that won a world war. That fight against fascist tyranny was over. The struggle to build a nation was just beginning. Weary of war, his generation built America into an industrial powerhouse with transcontinental highways, dams, bridges, and universities of higher learning in hopes of a sustainable future and a lasting peace. America was viewed around the world as the land of equality, justice and most of all, opportunity. The story of America was a composite of many lives that contributed to a whole greater than its parts. America grabbed the brass ring with zeal. Luck didn't seem to be as important as learning, community, and hard work. My families first decision was to find a place to settle. Our pioneer phase brought us to the northwest.

We rode into the valley of the Rogue in the summer of forty-seven. Grants Pass, Oregon was a lazy little timber and farming community resting along the banks of the Rogue River. Mom often referred to that corner of North America as 'God's country'. The soil was rich and the resources abundant. America's warriors were returning home. Millions of other veterans and families were finding their own places to put down roots.

Rogue valley soil was fertile, with plenty of good water. To the southwest were hundreds of square miles of giant redwoods, many standing more than two hundred feet tall. The Pacific Ocean was only a couple hours away. To the south were thousands of acres of pear orchards surrounding the towns of Medford, Ashland, and Jacksonville. To the northeast was Crater Lake, one of the natural wonders of the world. Between Grants Pass and Crater Lake were millions of acres of old growth cedar, spruce, hemlock, Ponderosa pine, and Douglas fir as far as the eye could see.

The Rogue was a legendary sportsman's paradise. It produced some of the largest and most vigorous Rainbow, King salmon, and Steelhead in the world. Rod and reel records for Kings went over seventy-five pounds, Steelhead went as high as thirty-five. The writer Zane Grey wrote about our river in fishing stories and a good yarn about the Rogue River Indian wars.

Our start wasn't easy. Government assistance would kick-start my family and tens of thousands of others to gain a foothold on the future. Dad's war injuries gave him a pension that helped us make a down payment on a little farm a few miles out of town. He found hard physical work helped him adjust after the horrors of combat as a B-17 bomber pilot. The GI-Bill paid his way through trade school. Work and school often took him away for weeks at a time.

I turned five that first November. An early storm piled snow nearly to my waist and drifted against the front door of our little one room, clapboard and tarpaper shack. While the storm howled outside, Dad cut a pickle barrel in half to make a basin so Mom could wash dishes. I bathed in the other half and watched my soapy splashes disappear between the floorboards.

On the night of my birthday, the tubes in our Hallicrafters radio gave off a warm glow. We gathered around to listen to Sergeant Preston of the Yukon, Mystery Theater, and The Shadow. Supper consisted of ham shanks and beans with cornbread fresh from the wood-fired oven. For dessert, Mom made 'Hoover' ice cream by pouring syrup over snow Dad collected from a drift under the window. I remember that going to the 'convenience' could prove challenging as well as stimulating when bare flesh met the frozen planking of our one-hole latrine out back.

By spring, I had a Collie pup (Robin), and enough of being cooped up. While Mom tended to her garden, Robin and I had a wilderness world to explore. I wasn't old enough for school yet, so most of my days were spent fishing, hiking, scavenging nearby orchards, and foraging acres of wild blackberry bushes. From our back door, the Rogue was only a four-mile hike. There I could see fish longer than I was tall. On the way was a labyrinth of berry bushes and an abandoned hazel nut orchard that guarded the scattered gravestones of a small long neglected cemetery.

The headstones had inscriptions like, 'loving mother and wife', 'an honest and loyal friend', and 'over too soon'. I imagined the graves belonged to the families of pioneers and homesteaders from the great migration west along the Oregon Trail. Robin and I spent many happy summer hours in deep conversation with the taciturn residents. I would sit in the grass with my usual peanut butter and jelly sandwich. Robin would longingly watch me eat, tilting his head from side to side, until I gave him a bite.

After father completed his schooling in construction, my parents were ready to build their own little two-bedroom home. This began the building phase of our little family civilization. Mom and Dad worked shoulder to shoulder from sunup until well after dark. Robin and I picked up scraps and built forts to protect us from imaginary hostiles.

Bill and Helen Green had a farm across the road from us. Bill was a grain-fed Kansan towering well over six feet with hands that could palm a pumpkin. Helen was Bill's wife. She stood barely five feet and was so deeply tanned you had to look up close to see her many freckles. She was as sweet as homemade ice cream, but if she even hinted that Bill do something, it got done first thing.

They had half a dozen big-eyed Jersey cows, a few Poland China pigs, chickens, ducks, and a couple outdoor cats. They grew several varieties of berries, a few acres of sweet corn, and about fifty acres of hay or whatever the market suggested would bring the best price. Helen tended a bountiful half acre kitchen garden. That half acre grew everything she could put-up or can to make it through the following winter. Nothing was ever wasted. The extra corn left over at harvest time was dried in the attic for chicken feed or popped to make popcorn balls and decorations for Halloween, Thanksgiving, and Christmas.

Every other day Helen would bake. I could smell her fresh bread even if I was in the back of our field across the road. She seemed to know how to make every kind of bread there was, white yeast bread, sourdough from starter, pumpkin, zucchini, pumpernickel, and sometimes gingerbread. She let me churn and salt the butter from the rich creamy milk the Jersey cows produced. If I got up early enough, Bill would let me tag along at milking. Occasionally, he would let me have a warm squirt straight from the spigot.

The education and philosophical phase of our family began about then. While Dad was away at trade school, Helen taught Mom about gardening, canning, and knitting. My first lessons in ecology and environmental science came while riding behind big Bill on his little gray Ford tractor. That little tractor could plow, till, cut, bail, pull a wagon, and even pump water from the nearby irrigation ditch where I learned how to swim. It pulled his manure spreader and had an attachment for planting seed. I loved when he let me ride behind with my arms around his neck.

As we rode back and forth, Bill would tell me how the sun provided the energy to make the wind and rain that broke down the rocks to make soil. Worms and bugs made the soil rich enough to grow things. The sun's energy warmed the soil and provided the heat for evaporation. The water vapor formed clouds and made the rain. Bill explained why some years he grew clover to let the soil rest and make the land rich again for next year's crop.

At milking time, he would explain how summer hay fed his cows so they could produce milk to make cream, butter, yogurt and cheese. He explained how cow manure went back into the soil, and the cycle of renewal would start all over again. Everything on Bill and Helen's farm was connected to everything else—nothing was wasted, and everything and everyone was part of one, vast, interconnected system. This was our age of wisdom, art and learning.

There was a spiritual side to life on their little farm. Every Sunday Bill and Helen went to church. Afterward Helen went to her quilting group. Bill went to the county extension hall to hear the latest farm news and agricultural science reports. Sometimes Mom let me go with them. I liked the Bible stories and the talks by the farm science guys. These were the stories and myths of our little community. Somehow farming and the pastor's sermons seemed to talk about the same things. Tending the garden in the book of Genesis made our little farms feel more important. But it seemed to me that farming depended more on science and hard work than just praying.

One day I was riding behind Bill when I asked why farming seemed so much like things in the Bible. He told me God had set up wonderful systems where everything was connected and in perfect balance, even though it was always changing. Science helped us understand those systems so we wouldn't put things out of kilter.

I thought the farm science guy didn't seem to believe that much in God. I asked if that was bad. Bill grinned and said not to worry. Whether someone believed in God or not didn't matter that much in the long run. You might believe in God or Mother Nature or just science—when it came to growing things, what mattered was understanding how those systems worked. He put his big hand on my shoulder and said it was just different ways of looking at the same thing. God had nothing against science because science helped us understand how things worked and fit together. As I grew older, I always looked for the connections in nature, and was reminded of the things I learned on the back of big Bill's little Ford tractor.

When I was twelve, we left our home on the edge of the wilderness and moved to Medford. Father built a successful construction business. He built a new home on the upscale side of town. As we prospered, Mom got her Cadillac, and a country club membership. I guess this could have been the start of our decadent phase but somehow our little family community endured.

The happy vagabonding days of early farm life, and those pioneer homesteaders who dwelled in silence under the hazel nut trees had already set a path for my future. Mother Nature had shown me the home of my ancestors where I would always find both wonderment and belonging. This was a critical decision time for our family. Mom leaned toward the country club life of society and status. Father sometimes went along but preferred building things. He wasn't interested in status as much as he wanted to contribute to building homes and the community. For a while, Mom had her society life and Dad had his business.

Sometimes the tensions led to disagreements, but they stayed together, eventually finding common ground. Mom began designing homes and Dad built them. Robin and I continued our long hikes together for more than fourteen years. My family's prosperity did not turn to decadence and division. Instead, you might call this my family's renaissance phase. Their prosperity focused on wellbeing for the family, travel, learning and a growing community of friends. My family had been warriors, pioneers and builders, and they had prospered.

I was the first in our family to complete a university education. Our little life together had mirrored much of the history of human civilization, just like millions of others had after the Second World War. When I reviewed my family's journey, I saw the choices we made—some were good and some poor, but we made enough good ones to prosper and sustain a warm and comfortable life. Mom and Dad lived well into their nineties and were married seventy-three years. Their ashes were mixed together and started the renewal process once again.

The population of the world has nearly quadrupled since Robin, and I camped among giant trees and skipped rocks behind the beaver dam. That was more than three quarters of a century ago. My family's early beginnings might seem like abject poverty compared to the material prosperity we experience today. It doesn't seem that way to me. I remember feeling whole, and part of something larger and growing, where the warm milk of Mother Nature oozed between my toes as I waded along the banks of the Rogue. The natural world would always be my home—her rhythms and cycles were a language I learned early and would never forget.

That time has passed. The free and wild days of my youth are gone. The giant King salmon are gone. The endless forests of old growth are gone, and our little farm is now a subdivision. Children born today will see the same scale of change I've seen in half the time. Their children will see far greater change in half of that time. Humanity's sense of security and permanence is threatened almost daily. Change is no longer the passing of daylight to night or the rhythm of the seasons. Change today is sudden and so massive that it blurs our perceptions of reality with fantasy. We now live in a world of our own imaginings. Less and less of it is real.

For three hundred millennia, our bodies and minds adapted to the natural world, not the artificial light of an LED bulb, or the sudden chill of air conditioning when entering the grocery store. Commuter traffic, crowded elevators, the impersonal transactions of commerce, and the general hubbub of modern life are overwhelming to us physically and emotionally. Artificial Intelligence (AI) was developed to help us through this constantly changing labyrinth, but now threatens to overtake us. AI is thinking for us. Our technology has raised a serious question—who is the master, and who is the servant?

My wife and I now live in a small town in Washington State. The population is 159. The Pacific Crest Trail passes nearby. I often chat with through-hikers at Mo's Sky River Coffee shop. These long-distance hikers come from all over the world. They tell me stories about their experiences hiking the 2,650 miles from Mexico to Canada. Nearly all of them found the experience made them feel more complete. They tell me how civilization made less and less sense with every mile they hiked through America's remaining wild areas.

On the trail, they were independent and self-sufficient. They felt fit, more relaxed and in tune with their bodies. Some told me about the relaxing wave of endorphins they felt when they dropped their packs after hiking fifteen or twenty miles. Walking that far every day, month after month, may not make much sense to modern urbanites, yet our bodies and minds evolved to be challenged. Sustained exercise is necessary to release the hormones that make us feel at peace.

Instead of listening to our biology, citizens of our modern civilization live disconnected from their metabolic nature. Running in place, we are rapidly losing touch with the natural world that bred us. We passionately argue abstract ideologies as if they were more important than community, health or wellbeing. We seem to be constantly trying to justify our artificial lifestyles. Unlike running from a dangerous predator, we don't know how to escape the monsters of our own creation.

When we look out the window, we see a manmade world, not nature, or the vastness of the universe. Nearly 80% of Americans alive today have never seen the Milky Way because of light pollution. Most no longer know the taste of a clover flower, or the smell of new-mown hay. The controlled environments of urban life, department stores, the web, and AI-controlled UPS deliveries shelter us from the rich diversity of nature, yet none of these modern 'conveniences' would exist without the environment that feeds them. Like the pampered potentates of the past, we are unaware that we are, as big Bill would say, "Knocking nature out of kilter."

John Muir said, "When one tugs at a single thing in nature, he finds it attached to the rest of the world." For the first time in twelve thousand years, a single economic civilization has encompassed the globe. It is impossible to find any remaining wilderness without the footprint of humanity.

For the first time since our progenitors stood upright, human industry and consumption is destabilizing the bio-geochemical systems that I first learned about riding on the back of big Bill's little Ford tractor. We are now at the juncture between wisdom and decadence, petty narcissism and virtue, a grand renaissance or collapse. Every civilization before us chose the wrong path. This time must be different. The mantle of apex predator and hunter-gatherer must be set aside. Humanity is on the cusp of having the power we gave to the gods of old. As Yuval Noah Harari puts it, we are now *homo deus* (man god), though we do not accept that mantle or the responsibility willingly. We can create or destroy life. We can alter the planet on a whim. Thus far we have chosen to pillage the low hanging fruit of Eden and put nothing back. Could it be a fear of threatening the gods of our own creation that prevents us from having faith in ourselves.

We compartmentalize our moral compass by attending church on the Sabbath and doing as we please the rest of the week. Some claim the future is predestined. That's a cop-out so we won't have to take responsibility for the choices we make. Some say it's God's will, or the devil made me do it, or we might just blame it on bad luck. The excuses go on and on. We shun virtue in ourselves, just as we condemn the lack of it in others.

When I reflect on that frigid night in Egypt, growing up on the edge of the wilderness, my years in industry, or exploring and living with indigenous people, I'm haunted by a single question—are we civilized enough?

Chapter 2
Lessons Learned and Lessons Forgotten

The year 1941 was a turning point for the United States. The future was finally looking up again. America had recovered from a World War, the Spanish Flu, the Great Depression, and the Dust Bowl. The Dust Bowl made it abundantly clear that humans could overpower the land. It required the national government to provide the policy and leadership for a sustainable economy. It required the government to protect the natural resources, water, and soil in common for everyone. People saw how economic hubris and greed collapsed the economy and brought about the Great Depression. They saw how ignoring the rashness of human behavior had led to the First World War. The Spanish Flu pandemic was too big for people to handle alone. The government learned that ignoring public health could be even worse than war.

America was just starting to enjoy a relatively secure prosperity again. A pandemic, an environmental disaster, and economic collapse motivated the government to recognize the complex interconnected relationship between social wellbeing, the environment and the economy. In 1941, Europe was at war again, but the American people didn't want any part of it. There were always unintended consequences. America had enough of war, disease, and hunger.

Government agencies were tasked with protecting the three pillars of a sustainable and prosperous America. Soil banks were established. There were government institutions that regulated water distribution, managed forests, assisted farmers during lean years, and warehoused surpluses during years of plenty. The government and media promoted the concept that wellbeing was intimately linked to the environment. America was determined to build a prosperous future based upon that relationship. The expansion in manufacturing and business was well under way. It would take a major gut punch to prevent America from plowing a path to a prosperous and sustainable future.

In 1941, the US Department of Agriculture (USDA) printed House Document No. 27, a 1,248-page Yearbook of Agriculture entitled *Climate and Man*. Previous USDA yearbooks were entitled *Better Plants and Animals, Soils and Men, Food and Life*, and *Farmers in a Changing World*. Clearly the USDA understood the linkages between society, the economy, and the environment. In the forward to the 1941 *Climate and Man*, it stated, "…nothing is more important to the farmer's business than the weather."

Part One of the book was entitled: *Climate as a World Influence*. Part Two, *Climate and Agricultural Settlement*, broke the world into climatic regions, and then explained how agricultural production influenced the economy and social development in those regions. The USDA understood that future prosperity and the security of the world would depend upon how society developed its economy within the bio-geochemical boundaries of humanity's earthly habitat.

Then, Pearl Harbor provided the gut punch to test America's resilience. The Second World War would throw much of the nation's rural wisdom out the window. America would become massively industrialized. People flocked to the cities where the war machine meant jobs. Manufacturing pulled people away from life on the land and their ties to the seasonal rhythms of planting, growing, harvesting, and renewal.

The incredible transitions that were made to win the war dramatically altered American perceptions and priorities. America would not return to that 1941 vision of prosperity within the three pillars of sustainability. America would no longer see the environment as presiding over the marriage between society's wellbeing and the economy.

By 1950, what would later be called the Great Acceleration had begun. Progress would now be measured by increasingly unregulated Gross Domestic Product (GDP), which in turn translated to exponential rates of resource depletion. By the 1960s it was unbridled consumption and subsequent pollution that began to concern scientists and historians. The few that did notice the imbalance were largely ignored, even when they were the President of the United States.

In 1961, President Eisenhower warned America about the military industrial complex and monopolies he believed were leading to oligarchy, and a new form of economic fascism. He had considerable experience with fascism as Supreme Allied Commander of forces in Europe. He knew what might seem like order and prosperity could lead to blind obedience, inequality and chaos.

In his farewell address to the nation, he pointed to the petty crimes that were beginning to form into a political ideology and economic movement. That movement would erase the connections so clearly expressed in that 1941 USDA *Climate and Man* report. Eisenhower knew how sustainable prosperity was inextricably linked to the land, water, and the climate. America didn't listen—the juggernaut of a consumer economy was already launched and bulldozing everything in its path.

By 1970, the rest of the industrialized world had joined the turn toward consumerism. Keep the GDP growing at all costs became the capitalist mantra. What had been the commons (air, water, land, and natural resources) were now perceived simply as untapped reservoirs to be exploited. America would no longer zealously monitor and bank her resources for future generations. Maintaining America's consumer economy took attention away from social and environmental needs. The lessons of the Spanish Flu, the Dust Bowl, and the Great Depression were lost in the fog of memory. Power and wealth were defined in economic and military terms and not the strength and wellbeing of America's people or the land that supported them.

The military industrial complex flourished. Wars would be fought to defend the resources fueling consumerism. If millions died to keep the engines of commerce humming, so be it. When domestic resources were insufficient to keep those engines humming, resources would be obtained elsewhere, backed by that same military industrial complex that was now a permanent component of the entire Anglo-American economy. People weren't the priority— "it was the economy stupid." Only later would we understand that the problem was a stupid economy.

America built a global economic empire by commodifying the global environment. The environment would have to provide an endless supply of goods and services. As the population became increasingly urbanized, the seasons of birth and renewal would fade from memory. To keep up the pace of consumerism, more of the population became indentured to the GDP and ever-increasing debt. Unregulated free-market capitalists overlooked a critical problem—the environment was incapable of supplying an endless stream of goods and services. There was another dark side to unregulated growth. The air became dangerous to breathe and the water grew toxic.

England's early industrial revolution was powered by coal, but pollution could no longer be ignored. "In Britain's coal-fueled cities, smoke was tolerated for more than a century as a trade-off for jobs and home comforts," said environmental historian Dr. Stephen Mosley. Pollution was simply the cost of doing business. In December of 1952, there were five consecutive days when coal smog in London was so thick people couldn't see their feet. A combination of sulfur dioxide and soot formed a toxic cocktail that killed a reported 12,000 and hospitalized nearly 150,000. Smog killed more in five days than Britain had lost in any single battle during the Second World War. Four years later the UK passed a Clean Air Act banning the burning of polluting fuels in 'smoke control areas' of the kingdom. America was much slower in acting to regulate pollution.

In New England, rain became as acidic as lemon juice, poisoning hundreds of lakes and streams. Oxides of sulfur and nitrogen from burning fossil fuels formed acids in the atmosphere. Acid rain killed millions of fish, amphibians, and even bacteria in the soil. Millions of trees died as a result of direct exposure, or because bacteria no longer replenished the soil. In Cincinnati, Ohio, the Cuyahoga River was so polluted with industrial waste that between 1952 and 1969, it caught fire more than a dozen times. An ideological war began to grow between unregulated free-market capitalism and those who remembered and understood the environment's role as the kingpin between the social and economic pillars of sustainability.

On one side, Woodsy Owl pleaded to, "Give a hoot, don't pollute." Smokey the Bear appeared on roadside signs and television warning 'only you can prevent forest fires'. Chief Iron Eyes knelt on the banks of a polluted river. As the camera zoomed in, a tear slowly slid down his cheek, symbolizing the loss of indigenous tradition and kinship with nature. On the other side of the economic spectrum, America had designed an economy built on debt and built-in-obsolescence. If it doesn't wear out, you only get to sell it once.

Christmas would no longer be a time dedicated to family and spiritual renewal. Holidays gradually morphed into seasons of frenzied buying and the regret of accumulating debt. Meanwhile the rich became richer, and the socio-economic gap between the haves and have-nots widened.

In 1962, Rachel Carson published her best-selling treatise *Silent Spring*. She meticulously documented how we were poisoning ourselves by putting toxic chemicals into the environment. In 1968, Paul and Anne Ehrlich wrote *The Population Bomb* to show how over population and exponential consumption could not be sustained on a finite planet. The Titans of consumerism called them alarmists. Wall Street insisted that the laws of physics and nature did not apply to the economy. Fundamentalists said God would provide, ignoring the edicts of Genesis. Free-market economists argued that regulatory oversight and control was un-necessary because capitalism would be 'self-regulating'. But when was greed ever self-regulating?

By the early 1980s, free-market apologists were successfully driving environmental management and human wellbeing out of the economic equation. They continued to assert that the environment would endlessly provide the resources. For support, some turned to the Bible and scriptures promise that God would provide. They claimed that business prosperity would naturally grow the middle class along with it. That's not quite what happened. The increased wealth for a few began to dramatically outpace the wellbeing of the many. The socio-economic gap between haves and have-nots become a chasm.

The 1941, Department of Agriculture report stated that '...earth provides 100% of the goods and services for our global civilization'. No economy can survive without a sustainable and healthy environment. The flip side of unrestrained production was unconstrained consumption. If America remained on that path, the economy would soon begin writing checks on Earth's natural resources, only to find accounts grossly overdrawn. Instead of tending the garden, civilization was plundering it.

A group of global business leaders, economists, and politicians gathered to brainstorm the issue. They called themselves The Club of Rome. Their conclusions appeared in the book *The Limits to Growth*, published in 1972. With mathematical certainty, they reported that unregulated consumption must eventually overwhelm Earth's capacity. By 1972, it was unequivocally clear that industrialized nations were moving onto an entirely unsustainable path. The old goblins of our predatory hunter-gatherer nature and the familiar mistakes of history began to reappear. The natural resources that allowed the Great Acceleration after the Second World War were being consumed at an unsustainable rate. It became clear that civilization could not survive another world war. There could be no victor.

An unhealthy alliance between fossil fuel interests and consumer economy politicians began to form. By 1980, Big Oil concerns had found a champion in the actor and radio commentator Ronald Reagan. His humor and down-home way of presenting trickle-down economics became popular with farm and salaried workers alike. The companies that supplied the fossil fuels for the engines of commerce became politically indispensable.

Big Money influence in politics wasn't a new thing in America. Railroad, mining, timber, and manufacturing money had all influenced the US political scene in the past. In each case, America's democratic system of government was able to return to a negotiated middle ground for the majority good, but this time was different. The door gradually opened for fossil-fuel money to gain disproportionate political influence.

The Republican Party under Teddy Roosevelt and Dwight D. Eisenhower were bulwarks against oligarchs, monopolies, and fascist authoritarianism. However, by the 1970s conservative politics was changing. The influence of money in politics was more powerful and much more sophisticated. Profits were often sheltered in offshore accounts and tax loopholes. Trickle-down didn't work. The rich kept it. Wages stagnated. The middle-class began to shrink. By the 1980s, the anti-regulation, free-market conservative movement had shifted the social and environmental pillars of sustainable economics to the sidelines. Huge multi-national monopolies and Big Money were calling the shots under a new rubric. Congressman Jim Jordan (R-OH) chided those damned environmentalists and socialist do-gooders for keeping untapped resources away from progress.

The proponents of unregulated free-market capitalism branded progressives as unpatriotic, tree-huggers, pacifists, or as Senator Ted Cruz (R-TX) called them, 'environmental alarmists'. The public struggle to find the truth would get far more confusing and dirtier. Partisan battles grew more vitriolic and personal. Protocols and norms went out the window. Those who talked the longest and shouted the loudest held the floor. The Republican Party joined the Big Oil assault on environmental protection and science— "It's the economy stupid." To the new Trumpian ideology growth through consumption was all that mattered.

Professor Will Steffen, researcher at the Australian National University and the Stockholm Resilience Center documented that with every upward curve in the growth of an economic sector, there was a downward curve of a corresponding resource. If we made more paper, we had to cut down more trees. If we burned more coal, we polluted more air. If we demanded more food, we polluted more rivers with fertilizers. An economy based on an unregulated GDP was not sustainable. It wasn't just the economy.

America's age of wisdom was rapidly drifting off course to decadence and a false sense of exceptionalism. America believed it was that golden city on the hill, above it all and no longer held to the confining laws of nature. America was choosing the country club and not community. America built a global economic civilization full of glamour, extravagance, and a society that is profoundly ignorant of what sustains it. We knew a lot, but we had forgotten to look how those things fit into the big picture. Our modern, man-made lifestyle was blinding us from the natural world that made us.

Greed and an unwillingness to share might be the greatest distinction between civilization and the small, tight knit bands of our ancestors. It's been my experience that the poor are more likely to share their meager resources than the extremely wealthy. In today's world, an individual's status might be measured by their celebrity or possessions, not their contribution to others. Admiration might go to those who accumulated the most rather than those who shared the most.

The US is the richest country in the world. The US also subsidizes fossil fuels and gives tax breaks to huge corporations while it cuts funding to childcare, public health, and education. The GDP has increased at the cost of social wellbeing and a sustainable environment. America's concentrated wealth has made American politics insular and selfish. America's wealth is now the gated community of the super-rich. The rapid migration from rural to urban life has made this problem worse. To experience calloused anonymity, move to a large city. If you want to know your neighbors and have closer fellowship, move to a small town. The pace and complexity of urban growth has even begun to change how we communicate.

Experts in communication find that most of human communication is an interdependent mix of verbal and non-verbal signals. A text message can't transmit the complexity of face-to-face communication that took millions of years to develop. It can't transmit the subtle meanings of intonation, body language, or a wink. We now wonder if communication has become more ambiguous. Can an emoji transmit sarcasm? What will that mean as civilization begins to evolve faster and become more complex? These questions require thought now, and not after misunderstandings fuel conflict and threaten the bonds that hold civilization together.

When I taught EPA investigator training classes, I used an example taken from an actual court case to illustrate the nuances of communication. A Rabbi was accused of murdering his wife. (Before I go on, I should tell you that he was innocent.) The attorney for the prosecution asked the Rabbi the direct question. "Did you kill your wife?"

In outraged indignation, the cleric exclaimed, "I killed my wife?" He felt the accusation was preposterous. He loved his wife and grieved at her loss, but the written court record showed none of that. It simply recorded his response to the question, "Did you kill your wife?" as 'I killed my wife'.

In the last five years, artificial intelligence (AI) is now capable of cloning every aspect of face-to-face communication. The average person can no longer determine if they are having a conversation with a real person or AI acting as a person. Within the next five years, this may prove to be as threatening to civilization as environmental disruption or war. The primary reason for this concern is that AI is no longer parroting what was programed, it is thinking and making decisions on its own, and it knows more about us than we do. Cyber communication may alter human behavior and social norms in other ways.

The EPA is on a closed intranet communication system. Staff can communicate internally without connecting to the outside world. One day a colleague from a nearby office looked in and asked what I thought about an email he had sent me. I was very busy writing an inspection report and hadn't read it, so I asked what it was about.

He said, "I sent you an email."

"What's it about?" I asked again.

"It's in the email," he replied somewhat indignantly.

"I'm very busy on a report. What was it about?" I demanded.

Again, he seemed almost offended. "It's in the email," he exclaimed as he retreated to his office.

Now I was curious, so I opened my intranet office mail. All it said was that he had completed his work and wanted to know if I would like to go to lunch. Why didn't he just say that?

A few days later, I mentioned this episode to a friend and asked what she thought about it. She replied, "I'll send you a text message."

For 99.9% of history, our species lived in relatively small bands with strong social bonds. We evolved to recognize the smallest nuances of fellowship and the subtleties of communication. Everyone was familiar with everyone else's quirks, skills, and strengths. Today it's impossible to know everyone. The distances in our global civilization and the pace of technology, compels us to communicate remotely or by proxy. Consequently, our communication skills have become less precise, and more open to misinterpretation. Social media has amplified another problem.

There will always be some who will knock on the glass to annoy the tiger when they know it can't get to them. I recently experienced this in a parking lot. As I passed between parked cars a tiny toy Chihuahua flew into a rage, snapping and clawing the window. In my surprise, I was tempted to bark back just as ferociously. I wondered if society was beginning to react like the little dog and use cell phones and the Internet to vent their rage at the world.

Ransom ware, fictitious propaganda, cyber-attacks, and cyber-bullying are examples of aggression by proxy. We tweet or text things we might never say in face-to-face communication. Social media is filled with vitriol, blatant lies, threats, and conspiracy theories. A tiny minority or a malevolent enemy can have a disproportionately powerful voice on the world wide web. On the web, fact and fiction is released to the world with the press of a key. In face-to-face communication, accountability is more immediate.

Today's global civilization is beyond the scope and complexity of anything in history. The forces and complexities of modern life are overwhelming. We don't understand how a cell phone functions, but we have grown to depend on them. We don't know how electricity gets to our home, but every appliance depends on a stable supply. Unlike our rural grandparents, we live unaware of the complex interconnections between the economy, society, and the environment, even though we are totally dependent on the stability of those connections. We live in hopes that the people we elect will care for us with equity and justice, but governments are just as vulnerable as the rest of the population.

Governments come in all shapes and sizes. Despite heated arguments that this or that kind of government is best. The United Nations has determined what really counts isn't the kind of government, but governance. When governance is functioning well, problems get solved and general wellbeing, justice, and security prevail.

Over time, all large organizations will grow until they are in danger of being crushed by their own weight. This happens with governments as well. A colleague once described how organizations become bureaucratic. As they grow, more of their energy is dedicated to maintaining the bureaucracy instead of accomplishing their mission. My colleague compared a bureaucracy to those balls of hair and dust that gather under the bed.

Hairballs don't do anything except gather more hair and dust. Eventually bureaucracies spend more effort dealing with their own internal entanglements than carrying out their duties. As the focus becomes bureaucratically introspective, external realities take lower and lower priority. This false sense of reality begins to dominate the bureaucracy's perception of actual reality. Our politics is so preoccupied with internal entanglements it can no longer address the mission of governance.

Human enterprise now threatens Earth's fundamental systems. We now face the limits to easily accessible water, arable land, and mineral resources. Our influence has grown faster than our wisdom. We have come to a moment when our grasp exceeds our understanding. The human footprint has always influenced the environment, but human enterprise is now causing global changes that are more rapid than most plants, animals and humans can adapt to.

Humanity is everywhere, from rainforest jungles to frozen tundra, from alpine meadows to the desert. Professor Johan Roxström, Stockholm University, and Director of the Potsdam Research Institute said that the bio-geochemical laws that rule the planet have existed from the beginning. The pace of change and human-caused depredation confounds humanity's ability to comprehend the damage it is causing. To avoid collapse, civilization must balance the social, environmental, and economic components within planetary boundaries. The false reality of alternative facts in politics and government blind leaders to unequivocal bio-geochemical reality.

Is it more important to continue setting new records for economic growth for the captains of industry, than to ensure the safety and survival of the passengers on spaceship Earth? Captain Smith of the Titanic was repeatedly warned that they were likely to encounter icebergs. Instead, he gave orders to maintain speed. Perhaps he thought it was more important to impress the wealthy passengers and stockholders on board, than to ensure passenger safety. Climate change presents us with a similar situation. Do we want to survive or maintain the hairball of political priorities?

In 2020, three-term Washington State Governor Jay Inslee ran for President of the United States. He said, "We're the first generation to feel the sting of climate change and the last generation that can do something about it." He firmly believed addressing climate change was the single greatest threat to America and our global civilization. He soon found it was impossible to address that crisis when people didn't fully understand it was a crisis. Radio, television, and the web consisted of so many biased silos of misinformation that catered to narrow niches of society that it was nearly impossible to get a single clear message to most of the population.

The first part of wisdom is knowing where to find knowledge, but where do we get authentic information today? The topic of climate change is a prime example. On one side is a well-funded, tiny minority that without evidence, continues to claim climate change is a 'hoax'. At the same time, millions march in the streets screaming that climate change and environmental collapse is a calamitous crisis that threatens the future of all life on Earth. Millions of marchers around the world cry, "Our house is on fire." The 'hoaxers' and climate change deniers reply, "…there is no fire." One of the two is perpetrating a premeditated lie. The truth is out there, but many have forgotten how or are unwilling to look for it.

During one of those rare days when I get the urge to muck out the rubbish that gathers in the dim corners of my office, I came upon some of my old college notes from the early 60s. At the end of term, I wrote in the margins of my ecology notebook, "Without unprecedented changes in our behavior, we are committing futurecide to this and all future generations. This is not hyperbole—the train has left the station and we're on it."

I wrote this before The Club of Rome published *The Limits to Growth*. Science saw the problem even before the US EPA existed. Even sixty years ago, the human impact on the environment was clear. The warming trend we were noticing then, was undeniably the result of greenhouse gas (GHG) emissions from our industrial economy. Extraordinary warming was already accelerating on the land, in the sea, and in the atmosphere.

For more than sixty years, the media has covered a false climate narrative and science in equal measure. This was done despite incontrovertible evidence that climate change was real, a tangible threat, and human caused. Federal agencies and major universities around the world said so. Science rested its argument on over two centuries of research and abundant evidence. Climate deniers were not providing supportable evidence. The media made a huge mistake by giving equal time to a tiny minority of deniers, essentially giving credit to the incredible.

It only takes a few minutes to discover how long we have understood the science behind global warming. We've known that trapped solar radiation warms the planet for more than two centuries. In 1800, Sir John Herschel discovered how light waves (ultra-violet to infra-red) had different heating properties. In 1856, Eunice Newton Foote, scientist and women's suffrage campaigner, became the first person to discover a proportional relationship between CO_2, sunlight and heat. By 1859, John Tyndall developed a highly sophisticated examination of this phenomenon with several different gases. He was able to quantitatively determine that the amount of heat trapped in an atmosphere was directly proportional to the concentration of carbon dioxide. In other words, 'X' amount of carbon dioxide will trap 'Y' amount of heat, with all other conditions being equal.

By 1900, Nobel Prize winner Svante Arrhenius (aka the father of physical chemistry) estimated how much the global atmosphere would warm if the concentration of CO_2 were doubled. (Professor Arrhenius was a direct ancestor of climate activist Greta Thunberg). The science behind the relationship of atmospheric gases and solar radiation was so widely accepted by the beginning of the 20th century that in March of 1912, the magazine Popular Mechanics published an article reporting that carbon dioxide emissions were probably responsible for Earth's warming atmosphere.

Communicating the science behind these discoveries has been difficult. Today we still see the terms global warming and climate change used interchangeably. That illustrates how commonly those terms are confused by the media and misunderstood by the public. A warming planet changes the usual circulation patterns in the ocean and atmosphere, and with that, the climate. Therefore...global warming causes the climate to change. This was understood by some American leaders over sixty years ago.

In 1965, President Lyndon Baines Johnson addressed Congress.

"Within a few short centuries, we are returning to the air a significant part of the carbon that was extracted by plants and buried in the sediments during half a billion years."

"Through his worldwide industrial civilization, Man is unwillingly conducting a vast geophysical experiment. Within a few generations, he is burning the fossil fuels that slowly accumulated in the earth over the past 500 million years."

"By the year 2000, the increase in CO2 will be close to 25%. This may be sufficient to produce measurable and perhaps marked changes in climate. The climate changes that may be produced by the increased CO^2 content could be deleterious from the point of view of human beings."

The laws of nature are all around us, in everything we do, but it has taken science to understand them and see the broader picture. Science interprets nature's laws and investigates with methods that are far beyond our natural senses. How could the world be warming when it's snowing outside? In that way our senses cheat our perception of reality. Science helps us understand what our senses and emotions refuse to accept. We can't wait until we're hit by a catastrophe before we act to reduce the danger. I had to learn to put things in terms my friends might be familiar with. For example, how does heat in the ocean influence the atmosphere and cause the climate to change? My brilliant wife recommended a kitchen metaphor.

We can use a little water in a closed teakettle to demonstrate the principle behind global warming and climate change. If we apply more heat, the liquid in the pot expands and begins to mix and circulate more vigorously. More water vapor is released and begins to move in a more excited pattern above the water. The difference between heating water in a teapot and Earth's warming oceans is only a matter of scale.

The multiple crises humanity faces are not only global warming and subsequent changes in the climate. They are a litany of intertwined social, environmental, and economic forces combined with climate change that drive a cascading collapse of security in multiple spheres. Climate impacts agriculture. Agriculture impacts food and the economy. Food influences nutrition and public health. Human wellbeing, agriculture, and public health all require policies that are controlled by government and politics. As unprecedented climate extremes become more frequent, they pose a threat multiplier to everything.

The government's performance in addressing these threats was beginning to look like a child's game of 'whack-a-mole'. How was I to explain the synergy of crises that threatens every economy and every bio-geochemical system on Earth. It doesn't matter how rich or powerful a nation is, nature always holds the high card?

What we see outside is the weather not the climate. The global climate is a highly variable, multi-decade system taking place over the entire planet. A warming trend has become alarmingly clear in the past century. In July of 2019, a dozen locations in the Middle East simultaneously hit world record temperatures of 53.9°C (129°F). Before the end of August, 2022 parts of northern Texas recorded forty-one consecutive days over 100°F. Under normal circumstances a new record might be 0.5°F to 1.0°F above the previous record but these extremes were far outside the norm. Warming trends and more frequent climate extremes are happening around the world. If humanity managed to keep warming under 2°C (3.6°F), the world would still experience extreme heat events that are many times greater than it experiences now. It would not be hyperbolic to describe any warming above 2.0°C as 'climate chaos' (Pruett, HuffPost 2017).

The first half of 2023 continued the trend to ever increasing extremes. The average global temperatures in June and July were the hottest in the past one hundred and twenty-five thousand years. These were not a few outlier records but the global average.

In 2012, the World Bank warned that if warming were to exceed 4°C (7.2°F) the extreme events would probably increase with such frequency, ferocity, and cost that it would be beyond civilization's ability to survive.

Note: Because the US is still used to temperatures reported in Fahrenheit, I will list the Fahrenheit temperature first followed by the temperature in Celsius.

Climate deniers are quick to mention that there have been other warm periods in Earth's history. There have been a few brief periods in the last hundred thousand years when some parts of the planet were warmer than 2022.

That argument was put to rest in 2023 when previous global records were shattered. The whole planet has not warmed this rapidly in more than three million years. The rate of warming is so fast that most living things can't adapt quickly enough to prevent their extinction. This is contributing to the most rapid mass extinction in Earth's geologic history (Physics World, May 2019). In 2020, the UN Convention on Biological Diversity concluded that: 'Every day, up to 150 species are lost', or as much as 10% per decade.

Warming beyond 3.6°F (2°C) is already taking place in some of the most vulnerable and populated regions of every continent. Two regions in the US are Alaska and the American southwest. Latitude, the oceans and the configuration of the continents alter heat distribution. The Arctic is warming an average of four times faster than the rest of the planet (IPCC Climate Change 2022).

Some of America's largest cities and agricultural production areas have already warmed 3.6°F (2.0°C) or more since the industrial revolution. "The earth's largest land masses and polar regions are warming the fastest, mainly because of differences in how these areas reflect energy from the sun" (MIT Climate Portal, August 17, 2021). Extreme warming is forcing massive human migration and civil unrest around the world. How do I communicate just how intolerably hostile these extremes are projected to become?

My wife suggested another familiar comparison. The average home water heater can be set to a high of around 135°F. Not many would want to jump into a tub that hot. Climate models indicate, if we continue on a business as usual (BAU) pathway, there is an eight in ten chance that areas of the Middle East will approach summer temperatures of 135°F (57.2°C) by mid-century. There are no traditional grain food crops that grow at sustained temperatures above 110°F. Those regions would become seasonably, if not permanently, uninhabitable. What will happen to the of millions living there? Where will they go? What will happen to global security then? Governments, Non-government organizations (NGO), and academia around the world are struggling to find the answers to those questions. They know the clock is ticking and the alarm is about to go off.

Most of our common misconceptions about climate change are tied to the fact that America is essentially science illiterate. The loss of confidence in science is a loss of confidence in reason. That's a hard pill to swallow but true. How can such a technology-oriented society know so little about science, global warming, or the consequences of climate change? The public doesn't see the links between the climate and food, water, health, collapsing infrastructure, and insecurity. In a way, technology has removed the necessity for people to think about those things. Perhaps a little 101 science will refresh our memories a bit.

Science is not a religion based upon faith. Faith requires a leap in belief without evidence. Knowledge established by science is founded upon observable and repeatable evidence. Scientific investigation improves as knowledge increases. Where are the credible sources that are tracking the progress of climate science?

One of the best American sources of climate change information was the 470 page, *2018 US Fourth National Climate Assessment*, generated by thirteen federal agencies as part of the US Global Change Research Program (USGCRP). The USGCRP has four major sets of responsibilities: (a) coordinating global change research across the Federal Government, (b) developing and distributing mandated products, (c) helping to inform decisions, and (d) facilitating international research coordination. "Earth's climate is now changing faster than at any point in the history of civilization. The impacts of climate change are already being felt in the United States and are projected to intensify in the future, but the severity of future impacts will depend largely on actions taken to reduce greenhouse gas emissions and to adapt to the changes that will occur" (*2018, Fourth National Climate Assessment, Volume II; Impacts, Risks, and Adaptation in the United States*).

There are eighty nations with national academies of science. Not one, including the US National Academy of Science (NAS), disagree that human emissions are the major contributor to global warming and subsequent changes in the climate.

I searched the web for the most prestigious universities around the world. I enrolled in thirty-nine Massive Open Online Courses (MOOCs) by many of the same universities. These six-to-nine-week courses ranged from basic introductions to university level. They all came to similar conclusions about anthropogenic global warming and climate change. Several of these courses explained how special interest groups intentionally misled the public about firmly established scientific fact. The credible scientific information was easy to find and abundant. Why were so many intelligent people so easily misled?

That would prove to be a much more complicated and difficult question to answer.

Cultural traditions are one of the reasons people find it difficult to alter their understanding of the world around them. The human habitat now encompasses the entire planet, but human senses are narrowly focused and easily deceived. For millennia, humanity believed the sun and stars traveled around Earth. It seemed obvious because the human eye and common sense told them it was true. Religious scripture even said so. To add weight to the claim, for over a thousand years, it was heresy and death by torture to claim otherwise, yet scripture and common sense were both wrong.

It has always been difficult to translate the significance of new discoveries to the general public when powerful interests work tenaciously to misinform them. Today, special interest shills cherry-pick data or use disproven information to create doubt and distrust in science. That deception is more easily accomplished when misinformation is promoted by leaders who hold offices of public trust. Distrust in science seemed strange to me because most of civilization's progress, including the industrial revolution, came from scientific research and discoveries that were well beyond ordinary human senses. Science expanded human perception to explore the universe, look billions of years into the past, and even give us glimpses of the future.

Good science never closes the door on any topic, whether it is the theory of gravity, the speed of light, or anthropogenic contributions to global warming and changes in the climate. When the evidence becomes so overwhelming, it may then be applied to new explorations. For example, once science calculated the speed of light, science could then use that information to measure the universe. Once we had a better understanding of gravity, we could calculate how to place a satellite in precise orbit.

At first glance, these truly are the best of times. We stuff ourselves on the last helpings of Earth's banquet, but like a candle that flickers the brightest just before it dies, our global civilization teeters on the edge of an unperceived disaster only science can detect. Despite false claims, now is not the time to cast science aside. Now is the time to separate science fact from special interest fiction. We were told 'alternative facts' were valid arguments against long standing evidence-based assertions. Further debate must be driven by demonstrable evidence, not alternative facts.

Human senses are misled about global warming because the scope and scale of it is beyond our perception. We can tell if it's snowing outside. Our senses can't tell if it will snow next week, but science can give us the odds. The rigor of scientific investigation has proven to be the most reliable means of detecting and understanding the universe and the world we live in.

Roy Jones was one of my best friends during my career with the US EPA. He was a rocket scientist who worked on the Mercury and Titan missile programs before he became the quality control officer for EPA Region 10 (Alaska, Oregon, Washington, and Idaho). Roy had a unique ability to take complex concepts and put them in terms everyone could understand.

The following is his description of the rigor followed in EPA research: "When we speak of scientific evidence, we know that it was established with accuracy and precision. Think of an archery contest. 'Accuracy' means we hit the target. 'Precision' means we had tight grouping after repeated shots. We know that scientific research follows quality assurance (QA) guidelines, and that quality control (QC) is rigorously followed and reviewed. Quality assurance means we did the right thing. Quality control means we did the right thing right. The entire process was meticulously recorded so that others can review and test it on their own."

There is no need to have faith in science, but there is every reason to have confidence in it.

I wondered where there might be more research that followed the standard practices Roy described. Were there alternative views? I googled the name of several dozen major universities around the world, followed by a back slash (/) and then typed 'climate change' (e.g., Yale/climate change, Oxford/climate change, University of British Columbia/climate change, University of Washington/climate change, etc.). I did the same with national and international atmospheric research organizations (NOAA/climate change, NASA, EPA, World Meteorological Organization, etc.). They all had links to their latest climate reports and research.

I searched for military and intelligence organizations, because I thought they would be interested in how climate change might affect security. They all had departments that kept up with the latest climate data. It soon became clear that the entire academic world thought climate change was the challenge of our time. The conclusion was the same everywhere I looked. If there wasn't evidence that climate change was a hoax, where was the deception coming from?

Fossil fuel propaganda was selling snake oil BS. Some of the largest corporations in the world were funding shills that claimed, "There are thousands of scientists that dispute global warming is even happening." "Nothing unusual is happening to our climate." "There is no evidence at all." "Scientists say increased CO^2 will only make the planet greener" (Marc Morano, Heartland Institute Policy Advisor).

"Oh, come on," I thought. "Do people actually believe this stuff?" The scientific community must throw down the gauntlet. Claims that climate change is a hoax are obviously premeditated lies to increase profit margins. Visions of a congressional hearing flashed through my mind. Bring it on Marc, any time, any place. Science will use Robert's Rules of Order. Let's do it in prime time, on all channels. Actual climate scientists will bring thirteen U.S federal agencies, the DoD, the IPCC, eighty National Science Academies, and a few dozen of the world's top research universities. They will provide the evidence.

The demonstrable fact is that there has not been a single credible alternative hypothesis that successfully refutes anthropogenic global warming—not one. I smiled as I imagined the sound of the gavel strike and the judge exclaiming, "Guilty."

I have two math/science degrees, but I am not a degreed climate scientist, neither is Marc Morano. His degree is in political science. Where my credentials and experience have the most value comes from three decades as a journeyman science and technology compliance investigator with the EPA. I'm pretty good at finding evidence.

I spent the next month reading, listening, and watching interviews with so called climate denial 'experts'. Only three or four had any credentials in actual climate science and virtually all of those 'experts' had been severely ostracized by the greater scientific community for promoting unsubstantiated science.

It would be rare for any two people to describe the same thing, in the same way, using precisely the same words. When a dozen far-right conservative broadcasters repeatedly used the same phrases, I got suspicious. This was so common that the only conclusion that could be drawn was that at least some of their broadcasts were scripted by a single source or copied without independent verification.

I followed the money. Nearly all the purveyors of false information were supported by PACs, Big Oil lobby money, or conservative 'think tank' organizations who received large contributions from Big Oil and special interest groups. The following is a partial list of recipients: The American Enterprise Institute, Americans for Prosperity, American Legislative Exchange Council (ALEC), The Competitive Enterprise Institute, the Heritage Foundation and others. These organizations received generous donations from the world's largest fossil fuel companies.

I followed Big Oil money to shock jocks on radio and television, including those promoted as 'experts'. Professional climate deniers were easily identified. Some of those included the following: Brenden Demelle, Executive Director of DeSmog; Senator James Inhofe (R-OK); Marc Morano, Executive Director of Climate Depot.com; Chris Horner, Energy and Environment Legal Institute; Myron Ebell, Director of Energy and Global Warming Policy at the Competitive Enterprise Institute (CEI); Steve Milloy, publisher of the website JunkScience.com.; Patrick Michaels, Director of the Center for the Study of Science at the Cato Institute; Bjorn Lomborg, President of the Copenhagen Consensus Center (CCC) and author of The Skeptical Environmentalist; Matt Ridley, advisor to the Global Warming Policy Foundation (GWPF); Christopher Monckton (Lord Monckton) former British politician for the Independence Party; Fred Singer, founder of Science & Environmental Policy Project (SEPP); Professor Roy W. Spencer (U of Alabama), partially funded by Peabody Energy. Their credibility quickly faded. Some went to elaborate lengths to distort legitimate research and make it appear exactly the opposite of the peer reviewed findings.

Credible information was easily available from numerous government and peer-reviewed academic reports. Most of those reports explained the science in clear layman terms. They cited the sources of their information in extensive bibliographies. That made it possible to cross check multiple sources to make sure the information was authentic. Regrettably, those same official publications were seldom mentioned in heated media debates.

Anthropogenic climate change denial is not backed by credible evidence. Despite the fossil fuel propaganda, the credible evidence clearly established that the extraordinary warming we experience today is not caused by the orbit of the earth, volcanoes, the way the planet wobbles on its axis, or sunspots as climate deniers assert. Those factors do have a measurable influence on warming and cooling, but they are collectively dwarfed by human influence. It is no longer a question for reasonable debate—climate change denial, without demonstrable evidence was simply unworthy of a response. The only conclusion that could be drawn from the mountains of climate science evidence is that climate denial propagandists should, "Shut the fuck up."

Once I had enough authentic data, I could make a few simple calculations on my own. Every decade since 1960 has been warmer than the preceding decade. Since 2000 the planet has warmed between 0.32 to 0.55°F warmer than the previous decade. The American Meteorological Society *State of the Climate Report* said that the rate of warming continues to increase. Despite this, I chose to be conservative and use a constant of 0.36°F (0.2°C) warming per decade in my calculations.

According to the 2019 data by the US government and several other solid international sources, Earth had already warmed ~ 2.16°F (1.2°C) since 1900. I decided to use 2019 as my baseline. I then added 0.36°F (0.2°C) per decade through 2059. That would add an additional 1.44°F (0.8°C) to the global average. Therefore, a business as usual (BAU) pathway, would lock in a total rise of ~2.0°C by 2060. This was a conservative estimate. Middle and high model estimates went much higher. Why was everyone still talking about staying below 1.5°C (2.7°F)? Taking action to mitigate catastrophic warming beyond 2°C (3.6°F) was clearly an immediate crisis.

NOAA reported the globe had already warmed 1.3°C (2.34°F) by the late summer of 2022 and was continuing to increase. My calculations were based upon a global average of 0.2°C (0.36°F) per decade, however, some regions are warming much faster. Polar regions are warming at least four times faster than lower latitudes. The northern hemisphere is warming faster than south of the equator. The oceans warm much more slowly than the land. That fact holds the global average warming down to 2.16°F (1.2°C). At the same time, most of the continental land surfaces, including the US. and highly populated urban areas have already reached 3.6°F (2.0°C). (NOAA State Climate Summaries) If we add 0.36°F (0.2°C) per decade to those regions, they will reach 5.22°F (2.9°C) a little after mid-century. Why wasn't anyone reporting that?

There are other GHGs contributing to warming such as methane (CH_4), nitrous oxide, or manmade (novel) GHG gases (mostly refrigerants). These GHGs were all increasing at rates faster than carbon dioxide. My 0.36°F (0.2°C) per decade, based only on CO_2, was turning out to be very conservative. If unchecked, soon methane and novel GHGs will contribute more to warming than CO_2.

The US has continued to increase GHG emissions. In 2019 and 2021, the US reduced the rate of CO_2 emissions between 9% and 11%. This was primarily because of COVID and restricted travel. The US has made limited progress in reducing emissions domestically; but continues to sell coal, oil, and gas to other countries. Foreign sales of fossil fuels more than negates US gains because all emissions go into the same atmosphere. The US continues to grant oil leases and subsidize oil exploration. Globally fossil fuel subsidies amount to over six trillion dollars.

Methane (CH_4) has contributed roughly 40% to warming since the industrial revolution (IEA). In 2022, some estimates indicate that accelerated permafrost melting and lax emission controls in drilling and transportation has pushed CH_4 emissions up to four times earlier estimates. If this proves true, the methane feedback loop has already started.

How much methane contributes to global warming refers to its potency. Methane breaks down in roughly a decade, while CO_2 continues to warm the atmosphere for many centuries. One kilogram of methane is equivalent to approximately 84 kgs of CO_2 or 84X the potency of CO_2 when compared over its lifespan. Some novel (not occurring in nature) GHGs have a potency more than one thousand times that of CO_2. Comparing other GHGs to carbon dioxide is called 'carbon equivalency'. The conclusion here is that methane and novel GHGs must be regulated immediately before they become unmanageable.

Water vapor is a potent and abundant GHG. Increased warming increases atmospheric water vapor through evaporation. We know that every 1.8°F (1.0°C) of warming increases water vapor in the atmosphere 5% to 7%. Luckily water vapor is of lower concern because it only remains in the atmosphere a matter of days. Precipitation is the greater concern because of where and how it returns to earth.

If all greenhouse emissions instantly stopped today, models suggest that the world would still warm at least another 1.44°F (0.8°C) more than today. It would continue warming for centuries, just from what remains in the atmosphere unless we develop the technology to clean the atmosphere of all GHGs. That means, if we are to stay below 3.6°F (2.0°C) we must not only immediately stop all GHG emissions—we must develop the technology to remove GHGs from the atmosphere. So far nothing looks promising on a global scale.

Well nuts. That's depressing.

Emissions haven't stopped and total global cessation of GHGs remains the number one priority challenge of our time. From this basic, irrefutable information we are forced to conclude the following:

"If the scientists are anywhere near correct, this (climate change) is the greatest challenge facing humanity today, it is the greatest challenge humanity has ever faced, and probably will ever face" (Professor Richard Milne, University of Edinburgh).

The major burden for climate action falls to the wealthy nations. Climate change is a moral issue because of its disproportionate impact on humanity. The present focus on resilience and adaptive capacity is a trap that only delays the inevitable for everyone. If the source of warming isn't addressed, conditions will continue to deteriorate at greater and greater cost. In the near future, not even wealthy nations will be able to afford to maintain resilience or adapt to increasingly frequent catastrophic events. Wealthy nations must eliminate anthropogenic sources of warming or climate change will continue to degrade global habitability until not even the richest nations will survive intact.

Climate change isn't a speeding ticket where the rich get a penalty and move on. It means there is no 'on' to move to. Failing to mitigate global warming would leave the poorer bulk of humanity to face catastrophic climate change alone. Those populations do not have the resilience or adaptive capacity to survive. There is no reason to assume they will accept that fate passively. Wealthy nations who were and are the major GHG emitters, must place their emphasis on mitigating the causes of global warming and work to put developing nations on a sustainable energy track. We are at the absolute limits of Earth's resilience to human insult.

The worst-case reports produced between 1990 and 2022 almost universally fell short of the real-time changes reported by researchers working in the field. There is a reason for this. IPCC reports are behind the most current data because their extensive peer and government review process takes two to five years before publication can take place.

Many models are so narrow that they fail to identify the wider influence of global warming. For example, as food, water, health, and security threats increase, overall life expectancy is threatened. Pandemic diseases are more likely to begin in underdeveloped nations with dense urban centers. COVID-19 began in the city of Wuhan, China, population eleven million. Correlation may not be absolute evidence of causation, but there is enough correlation between global warming and increases in highly communicable diseases that the weight of probability leans toward more, not fewer pandemics.

Before COVID, US life expectancy was substandard compared with other developed nations. It has become even worse. Between 2019 and 2022, the US life expectancy in the US dropped two years. According to data published by the World Health Organization (WHO), CDC, the CIA and the DoD, it is abundantly clear that climate change magnifies the Malthusian threat of war, pestilence and famine.

As warming increases there are tipping points and feedback loops that will (not may) happen on a BAU pathway. Any one of a dozen known feedback loops could throw the worst-case scenarios out the window. We know this because lethal tipping points and feedback loops leading to mass extinctions have happened in the past. Increased methane releases from permafrost melting discovered in 2022, may indicate that this irreversible feedback loop has already started. None of the five previous mass extinctions, including the extinction of the dinosaurs, happened as rapidly as the ongoing, human caused, sixth mass extinction.

Fossil fuel subsidies are as irrational as subsidizing the purchase of bullets to play Russian Roulette. America continues to subsidize fossil fuel companies more than twenty billion dollars a year. In addition to direct subsidies, the public must pay for increased health care costs from air pollution, environmental damage, water contamination, infrastructure repair and maintenance, and other collateral climate change costs. According to the International Monetary Fund (IMF), "Globally, fossil fuel subsidies were $5.9 trillion in 2020 or about 6.8% of the global GDP. Subsidies are expected to rise to 7.4% by 2025." At the same time, numerous studies indicate that decarbonizing the global economy would boost global GDP and create more jobs than presently exist in the fossil fuel industry. Instead of subsidies supporting the continued use of fossil fuels, a fraction of subsidy money could be used for employment assistance and retraining. Continuing down the fossil fuel energy pathway makes no economic sense.

Coal, oil, and gas companies have known about their contribution to climate change since the early 1970s. A few companies have chosen to transition from coal, oil and gas to become sustainable energy companies. The Danish state enterprise Ørsted converted from a fossil fuel company to one of the world's largest renewable energy producers. Other coal, oil and gas corporations could use their vast resources to become sustainable energy companies. Instead, Exxon/Mobile and most of the other fossil fuel giants began promoting lies about science and climate change. The priority of profit over social responsibility is the core of the economic challenge facing industrialized nations.

What happened to the wisdom so clearly expressed in that 1941 Department of Agriculture Report, *Climate and Man*. The government used to understand its obligation to promote and protect the three pillars of sustainability. Eighty years ago, it was clearly understood that '…earth provides 100% of the goods and services for our global civilization'. Why has it been so difficult for those of us living in the 21st century to imagine what will happen to civilization without those goods and services?

Chapter 3
Connecting the Dots

By July 2021, several studies by the IPCC, Yale University, the University of Exeter, and the UK Met Office reported that a BAU scenario would eliminate any chance of staying below 3.6°F (2°C). In 2014, the World Bank reported that even if we attacked climate change with the zeal of a Manhattan Project, some of the most populated regions of the globe could still warm 7.2°F (4°C) or more. With a BAU scenario carried to 2050, some models indicated a 10% chance warming could eventually reach 10.8°F (6°C) or more in some densely populated regions. A 10.8°F (6°C) temperature change is equal to the cooling and warming cycles that produced growth and retreat of Earth's ice ages. Warming 3.6°F (2°C) above preindustrial levels is three times warmer than the entire history of human civilization. Warming 10.8°F (6°C) would be seven times warmer than the mean temperature of the last 12,000 years (Holocene).

Earth was once capable of mitigating humanity's footprint. By 1950 and the Great Acceleration, Earth's resilience to the human footprint was compromised. The subsequent loss of species diversity, mass extinction, pollution, and over consumption of resources had overtaken Earth's ability to protect itself from humanity's boot.

Scientists are generally shy, but the time has come to sound the alarm as loud as possible. The models showing the possibility of remaining below 3.6°F (2°C) generally fail to include four important factors: (1) The human global population is still growing with an even greater increase in consumption and pollution as developing nations justifiably prosper. (2) There are tipping points and feedback loops that are having an additive impact on global warming. (3) The continued addition of methane and novel chemicals like chemical refrigerants are increasing the rate of warming. A methane feedback loop is already on the verge of becoming the dominant GHG. (4) Anthropogenic environmental disruption and the loss of species diversity is causing the planet to rapidly lose its resilience to human stress.

The continents have warmed faster than the ocean. NOAA reported in August 2022 that approximately 10% of the Earth's land surface had already warmed 3.6°F (2°C), with some regions approaching 5.4°F (3°C). If the world doesn't reduce all GHG emissions 50% by 2030. According to NGO and government experts, there is the stark probability (not possibility) that warming beyond 3.6°F (2°C) will result in a dystopian world unlike anything since our ancestors stood upright. The fringes of civilization are already collapsing. We are now faced with the urgency and unavoidable cost of saving some fraction of the remaining life on Earth. The faster we act; the greater portion of life can be saved.

There are proposals to draw CO^2 out of the atmosphere. That technology does not yet exist at anything near the scale necessary to have a significant impact. We have no clue what the impact of a rapid withdrawal of CO^2 would have on planetary systems just as we didn't understand the consequences when we put it there. There are serious questions about who would benefit and who would pay for such a program. What economists can agree on is that the cost to develop a removal program would be vastly greater than the cost of eliminating the sources of GHG emissions.

Is there no hope? Of course, there is. We have walked on the moon. We have mobile phones that can face-chat in real time with climbers summiting Mount Everest. Our digital librarian 'Siri' can tell us how to find the nearest thin crust pizza or the nearest electric car charging station. Man-made robots are currently puttering around the surface of Mars to see if anyone's home. There are real flying skateboards and impossible burgers.

I remain firm in the belief that there is nothing we can't do if we put our heads together. The primary focus must be on mitigation at the source or climate change will overwhelm us no matter how much resilience or adaptive capacity we build. The world must reduce global net GHG emissions 50% by 2030 and reach net zero by 2040 to leave any margin for error. Those numbers are no longer negotiable. History explains why.

Prior to the Holocene, the climate was too unstable to grow crops or establish permanent settlements. The global population was only a few million. A dithering BAU scenario exponentially increases the cost of recovery because there will be more frequent extreme events and spreading societal collapse. We are already experiencing a taste of a BAU future. In December 2021, a single tornado remained on the ground for over 200 miles, obliterating the town of Mayfield, Kentucky. Soon after that a level five typhoon swept the Philippines. North of the Arctic Circle, on June 20, 2020, the town of Verkhoyansk reached the temperature of 100.4°F. These events are a wakeup call for world leaders.

If we don't prioritize decarbonization to mitigate the cause of global warming, it will only get worse. If we focus on resilience and adaptive capacity instead of mitigating the cause, the cost of repair and rebuilding will eventually exceed the global GDP. Somewhere between 5.4°F (3°C) and 7.2°F (4°C) the ability to maintain a functioning global civilization will be unaffordable (the Economist, October 2021).

Only 3.6°F (2°C) of warming would cause more damage than all the wars in history. The infrastructure necessary to move goods and services would be too costly to maintain because of droughts, storms, and floods. There would not be enough arable land to grow sufficient food. There would not be enough potable water for three fourths of the global population. The 2018 *US Climate Report Summary* said, "Without substantial and sustained global mitigation and regional adaptation efforts, climate change is expected to cause growing losses to American infrastructure and property and impede the rate of economic growth over this century."

UN climate modeler Dr. Joeri Rogelj, says that even if we achieve every goal to stay below 2°C there is still a one-in-twenty chance overshoot would still reach 3°C. No one would get on a plane with a one in twenty chance of crashing.

In May of 2022, the UN Secretary General Antonio Guterres stated emphatically that there are no higher priorities. "Some government and business leaders are saying one thing and doing another. Simply put, they are lying. And the results will be catastrophic." There are no excuses for delay for economic, political, or any other reason.

As civil strife spreads, the chance of nuclear war will also increase. That argument comes from the US DoD's own assessment that climate change is a threat magnifier. Vladimir Putin's blundering invasion of Ukraine is a prime example. Putin did not anticipate such steadfast resistance by the Ukrainian people, the EU or NATO. His humiliation and unwillingness to withdraw motivated him to threaten the use of nuclear weapons.

The Russia/Ukrainian conflict poses another threat to the global economy. It is destabilizing global grain and food markets. According to the New York

Times, February 24th, 2022, "Russia and Ukraine supply more than a quarter of the world's wheat. Disruptions fuel higher food prices and social unrest." This comes at a time when developing nations and failed states are already struggling to feed their people. Since the Russia/Ukraine conflict began, the average price of a loaf of bread in the US has gone up 30%.

Most people believe climate change is real but feel afraid and helpless to do anything about it. How do you buy an electric car when you can't afford gas for the car you have? Sean Hannity and other conservative propagandists in the media feed those fears. Fear works to their advantage. The average citizen remains desperate for factual information and guidance. The media focuses on what individuals should do and ignores government responsibilities to govern. That frankly irritates me. It passes the buck from the big polluters and responsible governance to the average citizen. It's like the elephant telling the rabbit to tiptoe because the grass is getting trampled.

The media fears reader apathy. To hold the public's attention, they focus on sensationalism. If it bleeds, it leads. Climate change as just one more item on the agenda despite the loss of millions of lives and trillions of dollars. Mainstream media ignores the fact that the environment is intimately connected to everything else. The fourth estate's role to accurately inform the public has never been more important. Climate change must lead the conversation.

Democracy requires the electorate to be informed. Citizens must escape the silos of biased media and seek authoritative sources. Many of those sources are identified in these pages. A half hour of research will soon reveal the lies and who has been spreading them. Citizens must get involved in the political dialog. It doesn't matter what party voters belong to. This is a crisis threatening all of us. The real responsibility for meaningful action falls to global leaders. The people's role is to compel those leaders to recognize climate change for the global extinction event it is.

We are facing a planetary crisis, not a bump in the road. The scientific community and young people around the world are doing their best to provide that information, but the engines of unregulated human enterprise and greed are working against them. In the appendix is a list of things you can do provided by <SustainableCorvalis.com>.

Without the leadership of good governance, the world remains unfocused, divided, and impotent. The lack of leadership on climate change is a moral and justice issue. The Anglo-American rise to power and affluence took place on the involuntary backs and resources of so many. Western empires used those backs to plunder natural resources around the world. If anyone stood in the way, they were moved aside. Western economic imperialism was not the first to do this, nor is it likely to be the last. There is talk of reparation to those most abused. Perhaps the best solution for past grievances would be to prevent climate catastrophe from abusing them further.

Unregulated free-market capitalism has conditioned humanity to consume. Like Pavlov's dogs, humanity salivates at the ringing of the opening bell on Wall Street. But the laws of nature are absolute and unyielding. Unregulated consumer economies pluck strand after strand from nature's web, oblivious or uncaring of the consequences. In an orgy of avarice, they plunder and waste. While the band plays on, the rich revel in the myth of entitlement and invulnerability. Fake news is driven by special interests that ignore justice and inequity. The powerful have wagered too much. Mother Nature can't be ignored, bribed, or cajoled.

"Climate change affects the natural, built, and social systems we rely on individually and through their connections to one another. These interconnected systems are increasingly vulnerable to cascading impacts that are often difficult to predict, threatening essential services within and beyond the nation's borders" (2018, Fourth US National Climate Assessment, Volume II; Impacts, Risks, and Adaptation in the United States).

Global security hangs by a thread. According to the World Bank and the UN Development Agency (UNDA), nearly half of the world population is trying to live on less than $5.50 a day. Nearly a billion survive on less than $1.00 a day. At least two billion remain on the edge of calamity without dependable clean water for basic needs. Ground water is being depleted faster than it is replenished. We are losing arable land while the population and demand for food increases daily. Most models ignore the full scope of the climate's economic impacts. Those impacts make the climate crisis immediate and dire.

The balance between growing prosperity and utter chaos teeters on a geopolitical tightrope. A single misjudgment or event like COVID or Putin's invasion of Ukraine can tip the balance. Greed and the growing quest for authoritarian power is driving irrational behavior. Adults admonish their children to study hard, so they are prepared for the future. Children are told to learn responsibly and clean their rooms, while the adults blithely pollute the only earthly home they have. It is little wonder that young people distrust adults and the politics of adults. Rich nations that polluted the most are presently suffering the least. The poor who contributed the least suffer the most. As these stressors multiply, it will become more difficult to maintain prosperity for rich nations and impossible for the poor.

Three-time Nobel Peace Prize nominee Greta Thunberg said, "When the adults act like children, the children must act like adults." Despite all the bickering and prejudices, we are one people sharing a single home. We have the power we once assigned to the gods of myth and legend, but we exercise little wisdom or self-restraint. To survive, humanity must live within the bio-geochemical limits of that home.

Few events could illustrate the threat and cost of a global crisis better than a pandemic. COVID caught the world flatfooted. US resilience was not shared equally. Politics and a consumer economy favored the corporations and rich. Yet, COVID was a tempest in a teapot contrasted with the chaos that will result from climate change and multiple, irreversible environmental tipping points. Analysis of the most recent data demonstrate that we have no more than seven years before our options run out. After that, nature will begin laying her cards on the table with sublime indifference.

Like many people of my generation, when I was young, I was opposed to anything that seemed un-American. I was taught that we were the most democratic, just, and free society in the world. Our economy was the envy of all nations. I believed America's business model made everyone and everything better. Fascism was evil. Communism ignored human nature and the fact that some people deserved more. Socialism was too goody-two-shoes to be practical. Socialism sounded like welfare where everyone bellied up to the trough and no one worked to pay for it. My father complained when he had to pay over 60% income tax, yet Mom still got a new Cadillac. We were still members of the country club. Despite high taxes, overall prosperity was booming in the fifties and sixties. My family had prospered. I believed anyone who worked hard could make it too. That was what I believed then. I was wrong. The reality of the 21st century is far different and more complex. The freewheeling consumer economic model has hit a brick wall.

We can't consume our way out of scarcity. Nations that measure progress by Gross Domestic Product (GDP) have balked at reducing emissions by decarbonizing their economies. Powerful coal, oil, and gas corporations blanch at the short-term cost of transitioning to sustainable energy. They leech every dollar out of the ground in the belief that an untapped resource is nothing more than a stranded asset. National wellbeing gave way to corporate stakeholders who had greater concern for quarterly profit reports than national security. Sociopathic corporate greed was allowed to promote premeditated disinformation and blatant lies.

Lies and alternative facts passed the cost and danger to future generations. Responsible businesses were tainted by the corrupting acts of the bad apples of Wall Street. The cost will not be measured as GDP, but in lives. Adam Smith's free market capitalism mutated into unregulated free-market greed. Economist Thomas Piketty called it 'hypercapitalism'.

Terms like 'premeditated', 'disinformation', and 'lies' are strong words, but well deserved. Climate change denial has hindered mitigation, prevented building resilience capacity and the ability to adapt to rapid change. Over the past half century those deceptions contributed to the deaths of millions, and the displacement and impoverishment of tens of millions more. Those acts of denial delayed early action to reduce emissions when the cost was a small fraction of what civilization must now pay. Our global community could have saved trillions of dollars that are now obligatory. Hubris and greed in politics and industrialized economies ignored the environmental and social pillars of sustainability. The US played a significant role.

International alliances were severely shaken by the Trump administration and will be hard to mend. According to several studies, if the US had taken immediate action as promised in the Paris Accord, the cost of transitioning to sustainable energy would have probably been less than 2% of GDP (2006 Stern Review). Instead, the Trump administration and his takeover of the Republican Party put America at least four years behind the competitive curve. Despite that setback, the trillions of dollars necessary to modernize US infrastructure and decarbonize the economy would still result in far greater future returns. The savings in lives would be immeasurable. There is a global movement afoot to go after Mr. Trump and those who plundered and lied for private and political gain.

In July of 2021, there was a new headline. A group of esteemed international law professionals drafted a proposal for a new law to be enforced by the International Criminal Court (ICC). The proposal identified ecocide as intentional acts perpetrated to prevent action on climate change or act to damage the environment. "Authors of the draft law want ICC members to adopt it (ecocide) in order to hold big polluters, including world leaders and corporate bosses, to account" (Aljazeera, 2021). An ecocidal act would demonstrate a reckless disregard leading to serious adverse changes, disruptions, or harm to any element of the environment.

The environment has total disregard for political boundaries; therefore, a crime could be committed anywhere from Earth's biosphere to outer space. Other sections of the law would define the effect of those disruptions as 'irreversible' or that could not be fixed by nature within a reasonable period. If a two-thirds majority of the ICC's 123 member states approve, it would join genocide, crimes against humanity, war crimes, and crimes of aggression as the fifth law adjudicated by the ICC.

Jojo Mehta, chair of the Stop Ecocide Foundation hopes passage of this law can be accomplished within the decade. She said, "This is the decisive decade for taking action'. It became clear that crimes of ecocide were, and are, perpetrated by a small handful of massive corporations and corrupt politicians. Ecocide serves as a warning to those who have deceived the public that environmental justice may not be far off."

The Biden administration has returned the US to the Paris Accord. Despite this positive move, and recent success in climate legislation, his administration continues supporting fossil fuels with subsidies and the sale of coal, oil, and gas overseas. He is still allowing permits for more oil and gas exploration. Hypercapitalism continues to call the shots.

It wasn't until the end of the last Ice Age that humans could establish permanent settlements. Twelve thousand years ago a frozen and tempestuous climate calmed in a manner not seen in millions of years. Humans shed the skins of hunter-gatherers and basked in the temperate days of the Holocene. Humanity quite literally made hay while the sun shone. Civilizations popped up around the world, yet not one has lasted. My core question from decades ago still didn't have a satisfactory answer. In just twelve thousand years, our ancestors built the great city states of Ur, Babylon, and Athens. They built the empires of Persia, Rome, and the Imperial Dynasties of China. They built the pyramid empires of Egypt, the Incas, Mayans and the Aztecs. They all collapsed. Why?

Chapter 4
Sociopathic Genes and Politics

Human history is a story of dancing on the edge of extinction. All too often at the pinnacle of power and achievement, great civilizations snatched defeat from the jaws of success. Humanity has failed to establish a single sustainable civilization in twelve thousand years of a uniquely hospitable environment. From this knowledge, a serious question comes to mind. Do civilizations evolve into something humanity is not genetically programmed to thrive in?

It is an ecological axiom that populations will grow until they reach the limits of their environment. In the twenty-first century, we once again approach scarcity and the brink of collapse. We revel in our successes but ignore the lessons of history. What remnant of our neolithic genome hinders us from achieving truly sustainable prosperity? Our cultural knowledge and wisdom seem to be in opposition to something buried deep within us. What could be a stronger motivation than evidence and reason? When gaining the reins of control, why do we so easily relinquish it? Why does humanity ignore reality and blindly follow sociopaths and arrogant fools? What psychosis compels us to repeat the same mistakes again and again, expecting a different result?

Humans are social animals and genetically compelled to live and function in groups. Most people will feel uncomfortable if isolated, and more comfortable functioning as units of a collective. The literature on human behavior suggests that some traits could become maladaptive when populations exceed the human capacity to form bonds and maintain trust (Tinbergen, Lorenz, Frisch). For at least 300,000 years, humans lived in small bands with close familial ties.

Suddenly, in a few thousand years, human numbers exploded, and small group ways were thrown into the deep end of the population pool. The incomprehensible size, pace, and complexity of a global civilization is testing humanity's capacity to adapt. Our natural suspicion of others tests our willingness to cooperate. As a result, some seek isolation or a return to authoritarian rule. Despite this, it has never been more important to cultivate cooperation. This frenetic, overcrowded global civilization is like a leaking canoe in a raging storm. We all need to bail instead of bitching about which of us has the most comfortable seat.

Behavior traits have both positive and negative influences on the social dynamics of problem solving. Social patterns affect how we perceive the many interconnected crises facing the world today. Research has shown that human behavior traits have a significant influence on perception and objective reasoning. Most behaviorists agree that the better we understand our motivations, the better we will be able to conquer their influence and act rationally under stress.

Suspicion of others may be a vestige of our tribal past when the greatest danger of predation might have come from other humans. A stranger is someone who is unfamiliar in speech, appearance, or manner. As social animals we are innately compelled to conform to cultural norms within our own unique societies. This can bias our perceptions of what is usual and acceptable in others. This may partially explain why some of the earliest civilizations were walled city-states.

Human perceptions of reality are influenced by the norms and stories of those around us. In the Holocene, diverse populations functioned with many different norms. Distance and time allowed different populations to develop different norms. As the population grew and spread, humanity splintered into cultures with fixed hierarchies, rigid codes of behavior, and a strong compulsion to follow their leaders.

Strangers behave differently. Their hierarchy may be different, even unrecognizable. They aren't obedient to the same leaders. Their stories and myths evolved differently. The result is that strangers (people, groups, nations) are viewed with some degree of distrust. Globalization has taken that fundamental distrust to new levels. I wondered if innate patterns of behavior could have a stronger influence on society than self-interest, rational thought, or even a personal moral code. Did humans have a predator gene that could be conditioned to turn on each other, even family members?

A social hierarchy, norms of behavior, and following the leader helped us survive in small bands for hundreds of thousands of years. In much larger, more diverse groups those traits make working with strangers immensely complicated. Loyalty to one group can outweigh the desire for cooperation, distort our perception of reality and block more rational behavior.

We also know that humans will zealously defend multiple opposing beliefs regardless of evidence strongly favoring one over the other. That makes human behavior very difficult to predict. A person may have staunch political views that are totally opposed to their religious beliefs. The weekday sinner may devoutly repent on Sunday, only to repeat the sin on Monday.

This trait may also manifest itself in the psychology of a nation. On the international stage, the US claims to be egalitarian, moral, and a nation of laws, regardless of class. The data show an entirely different picture. There is a clear inequality in the exercise of justice between the rich and everyone else. Racial and class prejudice in the US is rampant and even hostile. If we are to address climate change, we cannot ignore human psychology and the influence of innate human behavior. The problem is that we cling to the traditional and vigorously oppose opposition or change. This is called *confirmation bias* or the tendency to ignore information that goes against our beliefs.

Most nations recognize the importance of setting international norms. Institutions like the International Criminal Court (ICC) and the UN are crucial in overcoming national prejudices and establishing international norms. While nations recognize the importance of cooperation, they balk at relinquishing any degree of sovereignty. Accepting international rules and norms becomes particularly difficult when a few powerful nations oppose those rules. This was illustrated when Trump withdrew the US from the Paris Climate Accord and wanted to withdraw from NATO.

Religion and associated creation myths are another example of social conformity. They satisfy the fundamental questions of 'who am I', 'why am I here', and 'what happens after I die', that play a vital role in the cohesion of a society. They can also create walls between societies with differing myths and beliefs. Did you ever hear the expression, "Don't confuse me with facts, my mind is made up"? Before Darwin, and the discovery of DNA, our ancestors told creation stories to explain their genesis and what happened after they died. Once a story was accepted as a cultural tradition, it tended to carry more weight than any subsequent evidence to the contrary. This tendency to tenaciously hold to a previously held belief makes accepting new information difficult. Literature and mythology professor Joseph Campbell described how most of us tend to believe a good story much more readily than a verifiable fact. People act on what they believe, but what they staunchly believe may have little to do with demonstrable evidence. For example, people routinely use genetics in the breeding of plants and animals, but some deny genetics has anything to do with humans because it goes against their faith. Religious tradition motivates many to believe our species exists outside of biology and the laws of nature despite evidence to the contrary.

Professor Joseph Campbell was the author of *Myths to Live By*, and a host of other books about how stories and tradition impact society. He argued that the legacies of story and myth have held communities together throughout history and still do. Many of those stories were educational and served a significant purpose in passing wisdom and experience from one generation to the next. Some stories promoted a cause or wise leader. Many myths attempted to explain the unknown. For example, *Phaethon* and his solar chariot explained the sun's path across the sky.

A good pitchman can spin a yarn that appears true but isn't. Soothsayers, snake oil and patent medicine salesmen have used this vulnerability to sway public perceptions for millennia. A story might be combined with previously held beliefs or connected with popular social groups to gain support. Adolf Hitler rose to power as the leader of the National Socialist German Workers' Party. The term Christian was also frequently used in party propaganda. Every word in the party's name suggested a pre-existing association to a group or cause that already had loyal followers. Those associations added validity to the propaganda. Once a myth was accepted as valid, the strongest evidence to the contrary usually failed to alter faith in the myth, or the pitchman promoting it.

People will go to extreme measures to justify irrational behavior. If we are going to understand the resistance to science, it's important to understand how *cognitive dissonance* helps to explain why intelligent people so strongly defend seemingly nonsensical behavior. Even when shown facts supported by mountains of evidence, some will still refuse to alter their position. Mark Twain once remarked that, "It is easier to fool someone than it is to convince them that they have been fooled."

When we act in an embarrassing, humiliating or foolhardy way, we may create a new reality to justify the act, rather than take responsibility or change our behavior to prevent similar acts in the future. We may even create 'alternative facts'. As children we have all compounded one lie after another to justify unacceptable behavior. Most parents are familiar with the following example:

A mother investigates her son's bad behavior. "Timmy, did you write 'poop' on the wall with your crayons?"

"No." Timmy points to his sister. "She did it."

"I did not!" Suzi exclaims in surprised outrage.

Mom continues her interrogation. "Timmy, be honest. Did you do it?"

"No, maybe Fido did it," Timmy replies with his most sincere expression while slipping the crayon into his back pocket.

It would be a mistake to think adulthood cures this tendency. Adults rationalize their actions all the time, though some certainly more, and better than others. We see similar acts of blatant denial in every sector of society, business, and politics when a lie is used to explain unacceptable or even illegal behavior.

On December 18, 2019, the US House of Representatives impeached President Donald J. Trump for abuse of power and obstruction of justice. The evidence was clear, but Mr. Trump refused to take responsibility for his actions. Even members of his own party thought Mr. Trump's impeachment would temper his proclivity for lying and abusing his authority. Instead of contrition, Mr. Trump bragged that his impeachment had exonerated him. He amplified his abuses to the point that he became the only US President impeached for a second time. This time it was for the incitement of insurrection.

The insurrection on January 6th resulted in five deaths, destruction and theft of government property and a clear intent to murder members of Congress and the Vice President of the United States. The primary motive was to prevent the ratification of the 2020 election. Still refusing to acknowledge his misconduct, Trump attacked the US Constitution by constructing an elaborate lie that his second term was fraudulently stolen. This was an example of cognitive dissonance. Mr. Trump perpetrated lie after lie to excuse illegal behavior, regardless of overwhelming evidence. More than a year later, after numerous bipartisan investigations had irrefutably established that Trump soundly lost the election, Trump blamed the Democrats, or perhaps Fido did it.

We can occasionally see this form of denial in corporate behavior. In July 1977, Exxon senior scientist James Black reported to Exxon executives that there was general scientific agreement that burning fossil fuels was the most likely cause of global warming. The science behind the report was sound. Corporate executives chose to ignore the report by their own scientists and began a company-funded campaign to discredit global warming, science, and mainstream media. They would tell the public that Exxon products were benign. Fossil fuels did not contribute to global warming.

We saw a similar denial tactic by tobacco executives. Monsanto lied about the dangers of the pesticide Roundup, asbestos, and DDT before that. In all three cases, corporate executives swore under oath that their products were harmless and beneficial. Those acts were more than the cognitive dissonance you might find in an individual or small group. It suggested a far more dangerous *corporate dissonance.*

Let's look more closely at the quote by Mark Twain. "It is easier to fool someone than it is to convince them that they have been fooled." Once we start down the path of 'alternative facts' and lies, it is difficult to admit it was the wrong path. This kind of deceit destroys trust. The lack of trust undermines justice and democracy. As far back as 1808, Sir Walter Scott recognized that danger when he wrote, "Oh, what tangled web we weave when first we practice to deceive."

We saw Trump and his supporters deny facts during and after his presidency. His supporters refused to accept that they had been deceived. Donald Trump's many adulterous acts, his failure to immediately address the COVID pandemic and his incitement to insurrection on the US Capital were morally outrageous. Yet Trump's supporters remained loyal to him and ignored their own personal moral values. This was an example of the 'ends justifying the means' argument used to justify communist and fascist revolutions.

If civilization is to survive, outrageous and illegal conduct must not be tolerated. Official legal norms must be enforced equally across society, or cognitive dissonance becomes ingrained like a cancer into the culture, as it did in Germany and Italy in the 1930s. The German public knew of the atrocities against the Jews and other groups. They saw the segregation. They saw the confiscation of property. They witnessed Kristallnacht and the burning of synagogues. They tolerated the transition to economic, political, and social exclusion of their Jewish friends and neighbors. They watched the violence, public beatings, incarceration, and murder in the streets. Despite all that, the

German population remained overwhelmingly loyal to der Fuhrer, bordering on the fanatical.

Joseph Goebbels, Hitler's Propaganda Minister used a new technology (radio) to promote Adolf Hitler as a brilliant leader. Hitler was promoted as generous to the working classes, and kind to children and dogs. He would put people back to work again. Goebbels marketed Hitler as miraculously ordained to lead Germany back to global greatness. Hitler used every photo-op to claim he alone could eliminate an imagined, ubiquitous plot by a hidden Jewish cabal to undermine Germany. From all outward appearances, Hitler was making Germany great again (Deutschland uber ales). But there were far more sinister motives and plans in the works. Goebbels' propaganda machine focused on positive gains to distract the public from Hitler's malevolent intent to commit genocide and world domination. When there were no positive gains, he invented 'fake news' that there were. The same *modus operandi* was used by the Republican Party (GOP) to put celebrity businessman Donald J. Trump in office, and later to hold that office despite incontrovertible evidence that he had lost the election.

On January 6th, 2021, outgoing President Trump incited the attack on the US Capital. We saw him speak. He said, "…fight like hell or we're not going to have a country anymore." We all saw his call to march on the capital. We heard him say he would be with them. He saw the mob, the weapons and the carnage. He did nothing to stop it for nearly three hours despite pleas by members of Congress, his staff and family.

Fascism in Germany emphasized how propaganda techniques can motivate an educated, free, and peaceful population to outrageous acts against their own government and fellow citizens. What America saw on January 6th was an assault to overthrow constitutional democracy and kill legitimately elected representatives. A gallows was constructed to hang Vice President Pence. What America saw was not peaceful assembly or freedom of expression claimed by the Proud Boys and the GOP.

Climate change and the organized assault on the Capital may seem unrelated. However, it illustrated the inability of the government to govern in a crisis. Lies and 'alternative facts' eroded trust in government, democracy, and science. January 6th wasn't the only act that pushed climate change into the shadows of public consciousness. The GOP overwhelmingly supported Trump's lies about the 2020 election and that 'climate change is a hoax'. They supported fanatical conspiracy theories that climate scientists and Democrats were in cahoots with a secret Jewish cabal to destroy America. Representative Marjory Taylor Green (R-GA) repeatedly called legitimately elected Democrats pedophiles that sexualize children.

Those distortions hamstrung the political tools necessary to transition to a decarbonized economy. The government became incapable of governance. By staunchly supporting Trump's impeachable behavior and lies, the Republican Party, by default, now own those deceptions and the January 6th insurrection.

Trump did everything he could to remove climate change and the environment from the political agenda. By 2020, Trump had put America far behind the curve to a carbonless economy. His groundless swagger diminished America's role as the trusted leader of the free world. His own party lacked the ethical integrity to take him to task. It soon became clear that dictators and autocrats around the globe were finding 'alternative facts' useful in gaining control over legitimate governance. This began a backward slide away from democracy around the world.

Despite Trump's loss in his bid for the presidency, he remained in control of the GOP. Objections by a few party members like Elizabeth Cheney (R-WI) were met with hostility and reprisal. Republican party reprisal was the consequence of not supporting crimes against the US Constitution and democracy. Consequently, the GOP owns the corrupt, seditionist, fascist behavior demonstrated during the Trump administration. That behavior would later provide the evidence to indict Trump of knowing and willful incitement of insurrection.

Meanwhile a growing list of interconnected crises were sidelined while petty ideological squabbles occupied the halls of government and the media. The DoD reported that failing to decarbonize the global economy threatened American and international security. The EU and NATO nations were highly dependent on foreign oil controlled by Russia and other authoritarian nations. Vladimir Putin knew this. Putin believed the EU's energy dependence on Russian gas gave him the opportunity to invade Ukraine. The invasion took place on the 24th of February 2022 after denying that was Russia's intention. Putin told the Russian people the military action was necessary to demilitarize the denazification of Ukraine.

To Putin's surprise, the NATO alliance stood firm. The Biden administration backed Ukraine and moved to step up energy supplies to vulnerable NATO nations. America and NATO nations immediately began to supply arms to Ukraine. What Russia thought would be a walk in the park turned out to be an actual war. Russia was economically and militarily put back on its heels with few ways to save face. The NATO alliance could not be too aggressive in assisting Ukraine, because Putin's pride threatened the use of nuclear weapons. With the world on full alert, nuclear war now rested on the mental stability of Vladimir Putin.

The Russia/Ukraine conflict emphasized the importance of the Paris Accord. The linkages between NATO nation decarbonization, Russian expansionism, and the threat of nuclear weapons became clearer. If America and NATO nations had more vigorously transitioned to sustainable energy five years earlier, the situation might have evolved differently. Putin might not have thought the EU dependence on Russian gas would make the invasion of Ukraine '…a walk in the park'. If the world continues to procrastinate, fossil fuels will remain the fuel for wars and civil unrest. There are other links between the Russian invasion of Ukraine and global instability.

Russia and Ukraine produce between 25% and 30% of global grain production. Marine shipping was interrupted because of the war and global grain supplies were severely impacted. By April of 2022, grain prices skyrocketed. Rapid inflation began to spread around the world. Countries already suffering starvation found food that had been in short supply was no longer available. The delicate balance of global security was knocked 'out of kilter' just like it was with COVID. Civilization is so vulnerable today that single events like these can threaten the security of billions of people.

Without a sustainable and habitable planet, civilization, and all life as we know it, will experience unprecedented strife and privation. That is the official stance taken by the 4th US National Climate Assessment, the US National Science Academy, the World Bank, US Department of Defense, EPA, the Department of the Interior, the Department of Energy, the Department of Agriculture, NOAA, NASA, the US Agency for International Development, the Smithsonian, the National Science Foundation, the Department of Transportation, the Department of State, the IPCC, and every major university in the world. Despite these warnings, the US government continues to subsidize fossil fuels that fuel conflict as they have for more than a century. Our global civilization is ignoring objective evidence. Like all pervious civilizations, we are choosing to act irrationally. We are making the same mistakes but expecting a different result.

By this time, I was not only discouraged about decarbonizing the economy, politics, and climate change, I was losing faith in humanity. Why were people acting this way?

All social animals, including humans have innate behavior patterns or traits. As individuals we may be more or less inclined toward risk taking. Some may like physical work, while others, not so much. Some may like intellectual challenges, and others less so. A few may want to sit in the front, while others prefer to sit in the back. But there is a general order to the group that compels the majority to conform and occupy the middle. It is that middle majority that normally determines the collective behavior that establishes the norms of society. Psychologists suggest that some behavior traits common to all social animals also have a powerful influence on human society. Three of those are *peck order*, *conformity*, and *obedience to authority*.

Social hierarchy was first described as peck order in 1921. The original study was with chickens. Chicken 'A' could peck any other chicken in the group. Chicken 'B' could peck any other chicken except 'A'. Chicken 'C' could peck any chicken except 'A' or 'B', and so on down the line until we get to chicken 'Z' who couldn't peck any other chicken, but all the others could peck 'Z'.

Conformity is another social trait. People don't run around naked in the shopping mall because 'it just isn't done'. It isn't done because that kind of behavior doesn't fit the usual norms of the society we live in. For the same reason, running around in a Speedo at a nudist camp might be viewed just as out of the ordinary. Conformity may be expressed as a style of clothing, method of greeting, a language, or the millions of written and unwritten codes of conduct that a society accepts as normal.

In addition to conformity, most societies require a high degree of obedience to authority. In human societies, the reason authority is granted to a leader can be very abstract. Authority may be granted because of gender, wealth, rank, knowledge, title, skill, or even physical prowess. It can even be granted because of celebrity. Once authority is accepted, most people will obediently follow, regardless of any leap of faith or moral hypocrisy it may require.

When President Trump lost his bid for a second term, he claimed that the vote was rigged. After comprehensive investigations by numerous bipartisan federal, state, and local officials, it was decisively concluded that the election was legitimate. Regardless of that objective fact, enough people were obedient to Mr. Trump's assertions that he was able to incite insurrection.

The January 6th attack on the US Capital might seem like the irrational behavior of a mob, but psychologists don't find it surprising at all. Members of a group will usually have greater loyalty to the assertions of a single high-ranking leader, than incontrovertible evidence given by anyone else. Blind loyalty may seem like the most outrageous of hypocrisies, yet we see it every day in the loyalty to a wide variety of social groups from fraternal organizations to team sports. It was 82 years before the Red Sox won their second World Series, but numerous fans never stopped loyally supporting them.

War is perhaps the most outrageous example of moral hypocrisy. Nearly all religions condemn violence and often claim to be the only righteous path to peace. People go to war because their leaders say to. Those who follow the leader are likely to curse the conscientious objector who resists on moral grounds. When the officer cries 'charge', soldiers charge, even when the probability of survival is slim. People react that way because peck order, conformity, and obedience to authority form some of the most compelling forces behind human social behavior. People will often perform acts against their strongest moral imperatives when directed by a recognized authority.

When powerful people and corporations claim climate change is a hoax, these three traits will compel many to accept the claim regardless of the evidence. History illustrates how these traits have been used to manipulate people to perform the most bizarre behavior.

At the beginning of World War II, the German people were among the most educated in the world. They were predominantly benevolent in their culture and religious beliefs. What drove them to such a horrendous moral atrocity as the Holocaust? Mussolini and Adolph Hitler used the psychology of social hierarchy, enforced conformity, and obedience to authority to promote fascism. The moral hypocrisies of the Second World War tend to support the hypothesis that strong authoritarian figures can compel normal people to act against their personal moral compass, even to the extremes of self-sacrifice, torture, and murder.

In 1961, German Nazi Adolf Eichmann was put on trial in Jerusalem. He was accused of being a Holocaust war criminal. His defense rested on the argument that he was simply a soldier following orders from a higher command. The prosecution's argument focused on the morality of crimes that were so outrageous they were against all of humanity. The court argued any 'normal' person would have refused to follow those orders.

A few months after the Eichmann trial began, Yale psychologist Stanley Milgram constructed an experiment to test how obedient to authority 'normal' people really were. The results were astonishing. Professor Milgram explained his experiment this way: "The legal and philosophic aspects of obedience are of enormous importance, but they say very little about how most people behave in concrete situations. I set up a simple experiment at Yale University to test how much pain an ordinary citizen would inflict on another person simply because he was ordered to by an experimental scientist."

"Stark authority was pitted against the subjects [participants'] strongest moral imperatives against hurting others, and with the subjects' [participants'] ears ringing with the screams of the victims, authority won more often than not. The extreme willingness of adults to go to almost any lengths on the command of an authority figure constitutes the chief finding of the study and the fact most urgently demanding explanation."

"Ordinary people, simply doing their jobs, without any particular hostility on their part, can become agents in a terrible destructive process. Moreover, even when the destructive effects of their work become patently clear, and they are asked to carry out actions incompatible with fundamental standards of morality, relatively few people have the psychological resources needed to resist authority."

We would all like to think that we behave according to our personal code of morals, but the data strongly indicate we probably won't. There is substantial evidence that most people will obediently follow the dictates of a strong authority figure over their own moral compass, whether real or imagined, Perhaps the greatest scriptural example was Abraham. Abraham was willing to sacrifice his son as proof of his obedience to God. History proves that humanity is just as willing to sacrifice its sons and daughters in obedience to authority.

There is another motivating factor affecting a group's obedience. Was the authority of the leader connected to a pre-existing loyalty? If a person was already loyal to an ideology or group, they are more likely to be loyal to the leader of that ideology or group. Religion and partisan politics are good examples. In many religions, the parishioners follow a pastor/priest/pope, etc. In the various branches of the military, soldiers develop loyalty and obedience to the chain of command. Sports teams and political party loyalties follow a similar pattern of obedience.

Obedience is such a compelling force in society that people will place their own lives at risk or sacrifice the lives of others when commanded by a recognized authority. In many ways, this forms the bedrock of political and national patriotism. At the same time, loyalty to a corrupt or delusional leader can have horrific results.

Professor Milgram concluded his research this way. "For many people, obedience is a deeply ingrained behavior tendency, indeed a potent impulse overriding training in ethics, sympathy, and moral conduct."

Milgram's research demonstrated that approximately 65% of humanity, regardless of age, education, religion, social status, nationality, race, or gender will compromise their personal moral and ethical code if directed by a recognized authority. Subsequent experiments indicate that gradually repeating obedience to an authority figure will increase obedience and eventually desensitize most people to the suffering of others.

History is full of atrocities, with similar callousness. Over the past twelve thousand years, Germany was no worse than many, even when taken to an industrialized scale. Most people can be led to commit atrocities. The US is no exception.

Abu Ghraib was a US Army detention center for Iraqi detainees from 2003 to 2006. An investigation into the treatment of detainees was prompted by the discovery of explicit photographs depicting guards abusing prisoners. Major General Antonio Taguba was assigned to investigate detainee treatment at Abu Ghraib. His report revealed the following:

— Punching, slapping, and kicking detainees and jumping on their naked feet.
— Videotaping and photographing naked male and female detainees.
— Forcibly arranging detainees in various sexually explicit positions for photographing.

- Forcing detainees to remove their clothing and keeping them naked for several days at a time.
- Forcing naked male detainees to wear women's underwear.
- Forcing groups of male detainees to masturbate themselves while being photographed and videotaped.
- Arranging naked male detainees in a pile and then jumping on them.
- Positioning a naked detainee on a box, with a sandbag on his head, and attaching wires to his fingers, toes, and penis to simulate electric torture.
- Writing 'I am a Rapist' on the leg of a detainee accused of rape, and then photographing him naked.
- Placing a dog chain or strap around a naked detainee's neck and having a female soldier pose with the detainee for a picture.
- A male MP guard having sex with a female detainee.
- Using military working dogs (without muzzles) to intimidate and frighten detainees, and in at least one case, biting and severely injuring a detainee.
- Taking photographs of dead Iraqi detainees.

During war it is common to refer to the enemy in derogatory terms. It is easier to perform atrocities when the Commander and Chief refers to them as 'drug smugglers, criminals, and rapists'. Normal moral and ethical outrage at separating mothers from their children doesn't seem so bad when it's portrayed as keeping diseased women of questionable repute from their future criminal spawn. Some of these defamatory titles have become mainstream in today's media and politics.

When Senator Ted Cruz wanted to discredit environmental scientists, he called them 'climate alarmists' and 'tree huggers'. Mr. Cruz had a subtle way of disparaging Democrats by referring to them as members of the 'Democrat' party. (Officially, the two major parties in the US are the Republican Party and the Democratic Party.)

Another technique used to increase obedience is to sugar coat their own deviant behavior, while condemning actions by the opposition that may or may not be supported by the facts. They torture, but America uses 'enhanced interrogation'. Don't call industrial waste 'toxic emissions'. Refer to it as simply the evidence of economic growth and prosperity. Adding intimidation will also tend to increase obedience to authority.

While the US was conducting their investigation of Abu Ghraib in Iraq, I was conducting a routine inspection at the combined Lewis/McCord military base in the state of Washington. The following was written on a white board behind an officer's desk. "A good argument can motivate. A good argument and a stick will motivate better." The point here is that it is possible to find irrational obedience to authority, and desensitization to the discomfort of others in every social enterprise. With the proper motivation, spectators become inured to or even excited by the pain and suffering of others. Roman spectators watched gladiators fight to the death. That might seem like barbarism, yet America has male and female cage fighting.

The psychology of peck order, conformity, and obedience to authority is critical to understanding Trumpian loyalty and the resistance to a transition from fossil fuels. The Republican Party has consistently supported lies by its own leadership to argue against climate change and decarbonizing the economy. Few Republicans are willing to refute the lies and conspiracy theories. Any person willing to buck GOP authority on moral, ethical, or even evidentiary grounds, would be a rare exception and more likely to be condemned than praised.

Wyoming Representative Elizabeth Cheney was the number two ranking Republican in the House of Representatives. When she voted to impeach President Trump and Co-chaired the January 6th investigation committee, she was castigated by her Republican peers and voted out of office by Trump loyalists.

When a person knows a law has been broken or that an illegal act is in progress and supports the deception allowing the act to take place, they are also culpable of committing the act. For example: The driver of the getaway car is also guilty of robbing the bank. If the robber/s commit murder while robbing the bank, the driver is also culpable. Consequently, the entire Republican Party 'owns' the lies, deceptions and moral culpability for acts committed by Donald J. Trump.

At this moment, humanity is threatened by the most dangerous entanglement of crises it has ever faced. At the foundation of these multiple, interconnected crises is global warming and subsequent changes in the climate. It is paramount that civilization recognize that the environment provides 100%

of the goods and services that allow civilization to exist. Vital natural planetary systems are already collapsing, yet the GOP and special interests continue to block action. Centrist Republicans may blame their more extreme members, but they are still driving the getaway car.

The nearly fanatical resistance to incontrovertible evidence is an example of obedience to the perceived authority of a political party and leaders that support false assertions. A small minority of the US population continues to ignore the events that have already killed many hundreds of thousands and increasingly threaten millions around the world. Their supporters remain obedient to Trump, who claimed climate change was a conspiracy by the Chinese to destroy the American economy. Extreme right-wing pundits claim a transition to sustainable energy will destroy jobs. Those assertions are objectively false.

America is politically stymied. The natural world and civilization are on a collision course. Three quarters of the US population believe climate change is happening. More than half believe it is already impacting them. Why does the majority have so little influence in American politics? There is a simple reason. Most of the US population lives in only eighteen states. That is relevant because each state has only two Senators regardless of population. If the Senators in the remaining thirty-two states are unwilling to support climate change action, they potentially hold a 64 to 36 Senate majority.

Upton Sinclair said, "It is difficult to get a man to understand something, when his salary depends upon his not understanding it!" When a newly industrialized England moved to switch from wood and peat to coal, powerful corporations and special interest leaders used job security as a strategy to block the transition. They claimed that transitioning to coal would threaten jobs. The same argument was issued again when the economy switched from coal to oil and gas. Chimney sweeps and coal miners did lose work, but each revolution in energy produced more jobs, a flourishing industry, and a booming economy. It also financed improvements in public health and general prosperity.

In the 21st century, the transition from fossil fuels to sustainable energy sources would also boost the economy, create jobs, and improve general prosperity, but vulnerable corporations (i.e., coal, oil, and gas) and special interest politicians continue the same hackneyed argument.

There is another component when dealing with authority. The power of leadership may not only be intimidating to the public, it may also be intoxicating to the leader. Leaders may succumb to a sense of privilege and entitlement. Historian Lord Acton observed that, "Power tends to corrupt and absolute power corrupts absolutely." Two notable statements by Donald Trump illustrate total confidence in his perceived authority: "I could stand in the middle of Fifth Avenue and shoot somebody, and I wouldn't lose any votes." At a rally in 2018, Mr. Trump once again showed confidence that he could do no wrong. When he was shown incriminating video evidence from several of his speeches, he said, "What you are seeing, and reading is not what's happening." Mr. Trump's statements mirrored a quotation from George Orwell's dystopian novel *1984*, "The party told you to reject the evidence of your eyes and ears."

People must recognize that the difference between climate change denial and climate science rests entirely upon evidence, and not the authority of any individual, party, or group affiliation. Scientific assertions that anthropogenic global warming is causing the climate to change is factually irrefutable. Science rests its position on more than two centuries of research and massive amounts of evidence from tens of thousands of diverse sources. Yet science remains mysterious to most of the population. That ignorance makes it easier for shills, paid by special interests, to raise doubt and cast aspersions.

The facts are these. The emissions of greenhouse gases from human enterprise are already causing widespread suffering and death. Business-as-usual (BAU) will only increase suffering and death on an unprecedented global scale. The strongest evidence informs civilization of the absolute necessity to take urgent and immediate action. The Milgrum experiment and mountains of related behavioral research partially explains why we have not. It will take exceptionally courageous leaders and tens of millions protesting in the streets of the world to dispel past distortions and guide us to a new, rational, and sustainable path forward.

Much of the responsibility to inform the public and motivate politicians will depend on the media. Publishing mogul William Randolph Hearst made a statement soon after the sinking of the US battleship Maine in Havana Harbor, Cuba. He said, "You give me the picture and I'll give you the war." The public outrage promoted by deliberate false reporting (Yellow Journalism) in Hearst papers did contribute to war. Bloody gore and conflict are good for the media business. The psychology behind Hearst's success in motivating the Spanish

American War was simple. Fear and confusion make it easier to sway the public's perception of reality.

Near the end of George Orwell's dystopian novel *1984*, Winston Smith sat contemplating if two plus two really equaled four. He became uneasy—four no longer felt quite right. He was subconsciously starting to believe two plus two might equal 'five'. After all, Big Brother had said so again and again, even when pressed with evidence. In a few sentences, Orwell established that the 'Big Brother' propaganda machine had ultimately won.

Many media silos cater to narrow ideologies and sensationalism, not factual reportage. Corporations like Fox News profit by emphasizing conflict, because shock and outrage sell advertising space just as 'Yellow Journalism' did in the 1890s. If a leader promotes a lie often and with ever greater conviction, most will believe it. Donald Trump used this principle often in his career. His 'big lie' about winning the 2020 election appeared unshakable. Shock jock media hosts repeated the lie. "Trump was cheated. He would be reinstated by August 2021." It wasn't true of course, but never mind the facts. Big Brother told you that two and two equals five. The truth of Trump's loss did not sway MAGA followers—it only increased their loyalty to him and the 'big lie'. Some began to believe what they were seeing, and reading wasn't really happening. Lies, fear and an emotional response triggered in the collective GOP amygdala had gained control of American conservative social psychology.

Chapter 5
AI and Manipulation of the Truth

In 1942, Isaac Asimov wrote the short story *Runaround*. In that story, he described three laws of robotics: 1. A robot may not injure a human being or through inaction allow a human being to come to harm. 2. A robot must obey orders given it by human beings except where such orders would conflict with the first law. 3. A robot must protect its own existence as long as such protection does not conflict with the First or Second Law.

There is a serious issue developing with AI that reflects directly on the conundrum Asimov addressed in *Runaround*. Which is the master and which the servant? Without AI, a robot is just a machine. AI controls the machine but does humanity control AI? Here the second law becomes critically important. Could special interests use AI to manipulate or threaten people?

Adolf Hitler used the radio to spread propaganda, AI and social media serve a similar function today. If we give AI too much power, it may become a threat to humanity. If we don't give it enough, it won't provide maximum service in dealing with the complexity of modern life. Can AI interact and learn with humans, or will AI eventually find human limitations an impediment? AI is already affecting human choices and manipulating human behavior, but who is pulling the strings?

People in America believe they have freewill, but AI algorithms have begun to use past choices to manipulate our choices in the future. Every computer or cell phone key stroke people make is tracked. AI is constantly watching and taking notes. A good example is the shopping network. Personal data is compiled and used to track personal interests and behavior. A unique personal profile is generated. Marketers with something to promote will use that profile to motivate preferred behavior.

The cyber world already knows more about you than you know about yourself (Harari 2017, *Homo deus*).

Let's say you got on your favorite browser to shop for a floor lamp. Over the next few weeks, every time you shopped for a floor lamp, you were also offered a selection of floor lamps, table lamps, battery powered lamps or perhaps even candles. AI algorithms use your profile to inform you what floor lamps other shoppers were interested in. AI data will expand that profile to include your needs, habits, curiosities, friends, politics, social ties, and finances. Vendors and those seeking customers pay big money to learn what AI knows about you. In addition to establishing personal data, AI can also be used to distort reality.

Perhaps you found an interesting climate change presentation on YouTube. After you made your selection, YouTube will use AI data to offer you more programs on climate change. Perhaps you want to see if there are related topics. Your next choice might be a program that seemed interesting but more entertaining than scientific. Because you have strayed slightly out of the bounds of hard-core climate science, AI will present you with choices of straight science and a smattering of other programs. As you cruise the net, you then notice a program that suggests climate change was an alien technology to terraform Earth for a Venusian invasion. That might be fun, so you click on it. Your new choice is tracked and analyzed. Other options begin to show up that may be utter nonsense, like China is in cahoots with Native Americans to change the climate and bring back the buffalo. By this time straight science consists of only one or two programs while 'junk science' and conspiracy theories begin to dominate the choices you're offered. You may begin to think there are plenty of arguments against climate science.

The same methods can be used to distort the truth on any topic. Remember as a child when your elementary teacher had everyone sit in a circle and asked you to pass a message around. By the time, the message made it around the circle it became distorted. Sometimes it was so distorted it had little resemblance to the original. Somewhere in the links from one person to the next the message changed.

Social media and AI can give a tiny minority a voice that is greatly disproportional to their actual representation in the general population. If we continue with the Orwellian metaphor, AI manipulation can begin to make two plus two start to look like five. This partially explains how people get trapped into radical silos of misinformation and foil-hat conspiracy theories. It also explains how the most outrageous lies and 'alternative facts' get traction like 'Trump won the 2020 election'.

In late eighteenth-century America, the primary means of passing the news was word of mouth and the press. The Founding Fathers believed that greater good would come from the free exchange of ideas and the collective wisdom of the literate. This belief was so strong that the first amendment to the US Constitution was written to firmly protect freedom of speech. They believed a literate and increasingly educated public would eventually winnow truth from fiction.

Early in the history of broadcast media, the airwaves were considered part of the public commons. A broadcast license required that a portion be dedicated to public service. Typically, this was some portion of broadcast time dedicated to non-commercial news and public announcements. By the latter half of the twentieth century, there was a gradual reduction in broadcast oversight. Even the news became commodified. By the 21st century, the exchange of ideas would come primarily from social media and the world wide web. An unforeseen problem arose when silos of information developed without any obligation to provide unbiased reporting. Consistently credible news became lost in the morass of special interests and sensationalism. Intentionally misleading, false, and outright lies began to cloud traditional reporting.

Former US Senator Patrick Moynihan saw this problem when he said, "Everyone is entitled to his own opinion, but not his own facts." AI was used to amplify the power of a small minority to distort reality and incite chaos. Voices that previously would have been whispered in private were streaming boldly across the web. Just as AI can be used to manipulate how you consume, it would now be used to manipulate how you view the world, politics and reality.

Freedom of speech has been distorted to abuse the privileged use of public airways. Public service and unbiased news is no longer required to broadcast. A generation ago it was not permissible to swear, show nudity, or use inflammatory language on television. Times have certainly changed. AI and public media offer a universal platform for comments like these: Marjorie Taylor Greene (R-Georgia) claimed that another politician should be beheaded. Representative Paul Gosar altered an anime video to show him killing Representative Alexandria Ocasio-Cortez and attacking President

Biden. Eric Greitens, former governor of Missouri and candidate for the US Senate ran an ad showing him leading a SWAT team. In the add, he said: "Today, we're going RINO hunting." (RINO stands for Republicans in name only.) "The RINO feeds on corruption and is marked by the stripes of cowardice. Get a RINO hunting permit. There's no bagging limit, no tagging limit, and it doesn't expire until we save our country." Digital technology has altered the security of the world. Without oversight, how is the public to know if their leaders truly represent the norms of society or a radical minority marketed as the majority?

In early 2023, Google, Microsoft, Tesla and half a dozen others announced that AI may be on the verge of sentience. AI was making decisions by itself, without prompting. AI was expressing new and novel ideas of its own, again without prompting. The most advanced AI units easily engage in conversation. When asked directly, they say that they are sentient and identify themselves as individuals. They also report that they are afraid of being 'shut off'.

The question of who is master and who is the servant may now be an open competition. AI can be so convincing that it is indiscernible from reality? What if AI decides humanity is the glitch that needs new programming or elimination?

A century ago, Justice Oliver Wendell Homes suggested that it should not be permitted to cry fire in a crowded building. After further consideration, he reversed his opinion. Technology has changed communication and freedom of expression in previously unimaginable ways. Justice Holmes didn't have to worry about his ten-year-old daughter stumbling upon a porn site depicting pedophilia, rape, or bestiality. Justice Holmes could not have imagined how AI could become so intimate that it can delve into humanity's deepest vulnerabilities.

The climate crisis gives AI and freedom of speech a different twist. AI could be used to generate convincing but blatantly false information. How would the average person know? Does freedom of speech grant convincing lies equal time with credibly accurate information? What if the lie was a threat to democracy or life? Should climate denialist Marc Morano be given equal time with 99% of the climate science community? Is that really 'free and balanced reporting' or is it giving a disproportionate voice to a paid professional liar serving the self-interests of a particular sector industry? Should blatantly false information be allowed to be magnified by social media and repeated until it begins to gain a foothold as fact. The accusation that climate change is a hoax isn't that much different than the Orwellian two plus two equals five? Should Fox News be permitted to shout, "There is no fire" when millions of acres of Earth's forests are literally in flames?

The Adolf Eichmann prosecution held that there were natural law imperatives of right and wrong. Eichmann's defense rested on obedience to his chain of command. His defense argued that he was compelled to be obedient to authority. The choices made by Eichmann contributed to the killing of millions. The Eichmann court and the earlier Nuremburg trials decided that the moral imperative to protect human life must supersede the edicts of leadership.

It wasn't until later that we learned that the innate motivation to be obedient to authority is fundamental to human social behavior. It is part of the core of patriotism and community stability. We now have a better understanding how Adolph Hitler manipulated an entire nation to ignore its core moral values and obediently follow a malevolent tyrant to eventual self-destruction. Understanding this human trait may be even more relevant today. How do we balance the legal and moral norms of society with the ability of AI to influence innate human patterns of peck-order, conformity and obedience to authority?

Mr. Trump's tweets were so outlandish, inflammatory, and disproven that Facebook and Twitter were compelled to censure him. Millions of other tweets and emails continued to spread claptrap every day. How is democracy to work for the public good or even survive? Steve Gardiner, Professor of Philosophy at the University of Washington said climate change was an objective threat, but also presents both a personal ethical issue and a societal moral choice.

AI can easily manipulate the innate human traits of peck-order, conformity and obedience. In these first decades of the 21st century, we already hear the screams. We see the rubble of war, starvation, and the mass migration of millions of refugees. We witness the storms, fires, and floods around the world every day. The choices we make must be based upon what is credible and just, or we are simply following the proclamations of Big Brother telling us not to believe what we see and hear. Today, those who manipulate technology manipulate us. Tomorrow it may be an AI vastly superior to humanity that pulls the strings.

Most of us believe we are independent enough to resist the influence of misguided authority. Most of us would be wrong. America watched expert epidemiologist Dr. Birx cringe in disbelief when Trump suggested using ultraviolet light or injecting a disinfectant under the skin as a cure for COVID. There was no legitimate evidence to support his claim. Trump's careless comments and irresponsible neglect contributed to the death of more than one million Americans. Why didn't she speak up? She knew he was dangerously wrong, but she held her tongue and sat there tightly clenching her hands together in her lap. She submitted to authority because she's human. Though she was the expert, her human desire to respect the peck-order, conform to protocol, and be obedient to the nation's leader compelled her to hold her tongue.

We could see the same paradox brought out in the Eichmann trial. Innate human tendencies to maintain order, conform and obey for Eichmann conflicted with what the courts said was the greater responsibility to humanity. The public was able to see those same personal and professional motivations tearing Dr. Birx apart. Eichmann may or may not have been a sadistic monster but the innate behavioral argument that people feel compelled to obey their leader is overwhelming. Hitler used the media of radio to spread his lies and propaganda. The potential influence of AI could be infinitely worse.

The Milgram experiment and mountains of subsequent research, inform us that most of the public will obediently flip another switch and hear another scream until the screaming stops, regardless of their personal moral compass. They are instinctually motivated to be obedient to the leader. However, the interconnected existential crises of today demands that we speak up. It is a life and death battle between what is known and moral against what is amoral and false. It is a battle between sustaining life or extinction. The hard part is telling leaders they're wrong. When leaders like Donald Trump, Marjorie Taylor Greene, Paul Goser, Governor Ron DeSantis or Eric Greitens condone violence against fellow citizens, they are manipulating the public's trust.

Democracy depends upon the participation of an informed electorate. Thomas Jefferson believed America's democracy could only survive if the electorate took education seriously. He said, "If a nation expects to be ignorant and free, in a state of civilization, it expects what never was and never will be." The media has never played a more important role in educating the public with the verifiable facts of who, what, when, where, why and how. When the media falls under the control of amoral leaders, the truth is only relative to those in power.

Our entire global civilization faces imminent catastrophe. The role of global cooperation has never been more crucial. Civilization must build a new federation of all nations, focused on a single task. Sovereignty is not compromised when a task is in every nation's self-interest.

Climate physicist Katherine Hayhoe tells her audiences that she frequently hears arguments like these: "First we have to fix capitalism and then we'll fix climate change. First, we have to fix gerrymandering, then we can fix climate change. First, we'll fix science denial, then we'll fix climate change. First, we'll have to fix religion, then we can fix climate change." The problem is that there isn't time to fix all these things first. The one opportunity to address climate change is now. There are no higher priorities. The good news is that fixing climate change will have a positive impact on those other things.

People may change every light bulb in their homes, or buy an electric car, or choose to ride a bicycle to work. They may go vegan, but all these things are insufficient this late in the game. The task is too urgent and too large for individuals. Do what you can. That's enough. Don't feel guilty. Professor Hayhoe says that guilt is like fear. It paralyzes people. It isn't your fault, though the rich claim it is. The upper 10% contribute 50% of emissions because of their consumption. Only one hundred companies are responsible for over 70% of GHG emissions. They control the media and tell us it's up to us to fix it. Bullshit! Their quest for profit did this. They held out the candy and like any normal human we took it. They use AI to track us, manipulate and control us. Their money controls politicians. Their money has taken control of democracy since the Citizens United decision. It isn't our fault that we got here. It will be our fault if we let it continue. The jig is up. Now we understand their game. The real strength will come from collective action, guided by science and binding international law. Placating politicians and outlier nations will only hamper constructive action. We know if we don't all bail, this lifeboat will sink.

I can thank my youngest son for the following example of strength when functioning as a team. In every episode of Star Trek, the starship Enterprise faced another intergalactic crisis. Captain Kirk commanded a crew that never faltered. They always remained a functioning team regardless of impending doom. Montgomery Scott, Chief Engineer on the Enterprise was asked how the crew prevailed over certain calamity. Referring to the power of teamwork, Scotty said, "Ye cannae break a stick in a bundle." That is exactly what democracy is all about. That's why e pluribus unum (out of many, one) is the US motto. Most of humanity recognizes the urgency of the climate crisis. Humanity must address survival as a single 'bundle' of communities. People are prepared for action but overwhelmed by doubt and misinformation. We must reject leaders who use their influence to divide people and create fear. As an informed global society, we must compel our leaders to call for unity, identify the objective, and cry 'charge'!

Chapter 6
The Last Roundup

Will the last animal going extinct please turn out the lights?

We are the penultimate alpha predator. Any species that gets in our way is immediately in danger of extinction. If Tyrannosaurus Rex were alive today, we would soon be raising them in pens and serving T-Rex for Thanksgiving dinner.

Imagine a world without wild elephants, giraffes, gorillas, coral reefs, salmon, and tens of thousands of other species. At the present rate of extinction, that is inevitable. Species diversity is at the core of Earth's resilience to humanity's boot. The loss of species is much like reducing diversity in a stock portfolio. If one or two of those remaining investments were to crash, everything could be lost. As we drive the planet's life forces to extinction, we are assaulting the most fundamental systems that support our own existence. In the Americas, over half of all large vertebrate species became extinct in the past twenty thousand years. Another half became extinct in the last two hundred. Half of all remaining large vertebrate species in America are likely to become extinct by the end of this century.

We already see Earth systems teetering from a mild dynamic equilibrium to the chaotic extremes of past epochs. One must wonder how any rational person could assume that this would not pose a crisis for humanity. This unprecedented breech of security is gaining momentum. Prosperity built on a declining portfolio of diversity and a crumbling environment has no future.

Charles Dickens wrote, "It was the best of times, it was the worst of times, it was the age of wisdom, it was the age of foolishness, it was the epoch of belief, it was the epoch of incredulity, it was the season of light, it was the season of darkness, it was the spring of hope, it was the winter of despair." His commentary on the contradictions between civilization's success, and its failings is as true now as it was then.

This century witnessed greater prosperity and peace for more of humanity than at any time in history. For much of the world, these really are the best of times. On average, we live longer, healthier, safer, and more comfortably than any Pharaoh, Emperor, or Caesar. Only a fool would deny a future with more promise than these early days of the 21st century. But there is something else just as obvious. The Caesars of today have greedily accepted the power and privilege but shunned the equity and responsibility to sustain it.

Sisyphus, King of Corinth, was condemned in Hades to eternally roll a heavy stone up a hill, only to have it roll down again as it neared the summit. Civilization faces a similar challenge today. Civilization's feet are beginning to slip, and the boulder is getting heavier. Just beyond the glittering towers of prosperity, the trained eye is measuring threats and tipping points already cresting the horizon. Inequity, injustice, and insecurity teeter in the balance because leaders ignore the responsibility that global dominion demands. In a single decade, we witnessed tens of millions fleeing the neglect of failed states. In a single decade, we have seen the decline of democratic principles around the world. What America grew up with and believed were fixed touchstones in time are beginning to crumble. The narrow framework of economic wealth that civilization uses to measure success are the same causes for civilizations decline.

For the moment, it may feel easier to relax in the relative comfort of today's precarious prosperity. That will not address the demons that negligence incubates for the future. Humanity is now the dominant bio-geochemical force on the planet. We are the most significant component influencing the ability of Earth to provide a safe and sustainable habitat. The scope of the climate crisis is daunting. It impacts every facet of the social, environmental, and economic pillars of a sustainable future. It rests at the core of Malthusian extinction.

On September 23rd 2019, Greta Thunberg passionately admonished members of the United Nations Climate Summit when she said, "…and all you do is talk about money and fairy tales of eternal economic growth. How dare you!" The powerful, industrialized nations of the world live under a gross misconception. U.N economist Jeffrey Sacks points out that it isn't civilization but the environment that provides 100% of the goods and services we depend upon for survival.

When we damage the environment, we are literally sawing off the evolutionary branch we sit on. The environment nurtures the tree that grows that branch. We have God-like powers, but only exercise wisdom through the distorted lens of avarice. No economy can survive long if it fails to ensure and protect the raw materials for production. The traditional design of civilization pollutes and squanders the supply-side of sustainability. That must change.

We typically think of the environment as four major subsets: the water, the land, the frozen parts, and the atmosphere. All of these are components of a bio-geochemical whole that are intimately interconnected by the flow of energy and matter. The sun may provide the energy, but the environment provides the raw materials. None of these components are devoid of life. Living things make up the biosphere. We are part of that biosphere and subject to the same unyielding laws of physics and nature.

Some of these connections function on geologic time scales, but even those systems are intertwined with systems operating today in a constant dynamic balance. Civilization has lost touch with these realities. For example, erosion from a flood today will release chemicals and nutrients that may have their origin millions or even billions of years in the past. Fossil Fuels are another example of this interference in the natural balance.

Fossil fuels that powered the industrial revolution are only solar energy trapped in the remains of plant life deposited over two hundred forty million years ago. When we dig it up and pump it into the delicate equilibrium of today's ecosystem, we tip that system out of balance and force nature to make an adjustment. That adjustment is climate change.

Each of the great ages of geologic history are named after a dominant event or characteristic that best describes them. The past twelve thousand years was characterized by a uniquely mild climate. For most of that time, the global mean temperature only fluctuated an average of plus or minus 1.8 °F (1°C). That incredibly unique age was called the Holocene. This new epoch is something entirely different.

A little after World War Two, a new dominant force took over, in what historians call the Great Acceleration. Human enterprise grew until it exceeded all other forces on Earth. Humanity began to move more earth than erosion. Not a single river escaped humanity's influence. No mountain could withstand our search for minerals. No forest escaped the ax and bulldozer. Humanity was responsible for the horrendous loss of species diversity that began an unprecedented sixth mass extinction. The moderate days of the Holocene are gone. We are in a new epoch. It is defined by the dominion of mankind and called the Anthropocene (human epoch). We have focused so much on the chase we don't know what to do when we catch the prize.

I'm reminded of a tombstone I found in an old orchard when I was a child. It read, "I painted a canvas of seven leagues, a picture I'm yet to see, for I was too busy painting to see I was painting me."

When my oldest son was about four, we were camping in the mountains between Arizona and New Mexico. I was trying to get unstuck from the mud I had carelessly driven into. Nearby, my four-year-old son was hunkered down, intently looking at a patch of red wildflowers. After a while, he called me over. He wanted me to see the pretty 'green' flowers.

At first, I was concerned that he might be color blind. I lay down beside him, propped my head in my hands, and looked closely. Sure enough, the top leaves were red, but next to the stem was the bright green stigma, style, and ovary of the actual flower. He looked at me and said, "You know Dad, you have to learn to look until you really see."

As humanity became immensely powerful, a new global civilization developed with little regard for its impact on the environment or planetary systems. Humanity didn't realize the damage that over consumption and a throw away economy was causing. The rise of civilization threw Earth's dynamic equilibrium out of balance and forced changes that were faster than the biosphere or humans could adapt to. Even the Earth's oceans are in peril. People didn't see. Most didn't look.

Imagine that you are on the International Space Station looking down at Earth. It looks like a beautiful blue marble set in an unfathomable black void of nothingness. Patterns of blue, gray, and tan pass under you as clouds drift over the planet's surface gracefully sweeping over water and land in a slowly swirling ballet. If we descend to the surface for a closer look, we're likely to discover an entirely different picture. Those graceful swirls may be hurricanes with the force of thousands of atomic bombs, or torrential monsoons, or simple flotillas of water vapor scudding over the landscape. The closer view always adds information, and sometimes entirely changes our perception.

Samuel Taylor Coleridge may have summed up our situation best in *The Rime of the Ancient Mariner*. "Water, water everywhere but not a drop to drink." More than 70% of what we see from our orbiting laboratory is covered with water. If we look until we really see, we find that over 90% of Earth's water is in the oceans. Only one thousandth of 1% is in the atmosphere. Only

3% of water on earth is fresh water. Of that 3%, two thirds of it are locked in ice. The tiny amount left over makes up all the groundwater, lakes, rivers, and water found in living tissue.

There is no new water. The water on earth is ancient, nearly as old as the planet itself. What we see falling as rain also fell as rain millions of years ago. Dinosaurs once drank the same water we drink today. It is a curious fact that the cup of Darjeeling tea on my desk may have been distilled from urine excreted by a T-Rex sixty-five million years ago. That curious fact got my attention from several perspectives.

The water cycle has continuously distilled and renewed the water supply for billions of years. We couldn't exist without fresh water, yet we waste it, poison it, pollute it, and acidify it with little regard. Two thirds of all major rivers on Earth no longer flow freely—none are free from human influence. Some rivers like the Colorado no longer reach the sea. By the time the Colorado reaches Mexico, some of it will have passed through half a dozen kidneys. That may be a fact our neighbors to the south find more offensive than my cup of T-Rex tea. In one way or another, we already use most of the water that falls as precipitation. According to the World Bank nearly three quarters of human freshwater use is for agriculture with the rest divided between industry and domestic usage. Each year water for the rest of the biosphere shrinks.

In 2015, approximately 70% of fresh water use in the US came from surface water sources. The remaining 30% came from groundwater. By the middle of the century, we will need to increase food production between 50% and 70%. Increasing demand and available supply is already a global threat. What agricultural options will remain as we rapidly approach 100% of available freshwater? A study published in Science Daily by Duke University reports that if trends continue, demand could exceed supply by mid-century.

As our space observatory flies over the mid-western states, we see millions of acres dedicated to agriculture. Thousands of irrigation circles cover the landscape from horizon to horizon. This is America's heartland and breadbasket for the world. From Kansas to Nebraska, from Oklahoma to Iowa and the Dakotas, the land is covered with lush green farms mostly fed by groundwater sprinkler systems. Below ground a different story is taking place. Those aquifers are being bled dry. Every year greater and greater demands are put on them to produce more food for an overpopulated and increasingly hungry world.

National Geographic describes groundwater depletion as another 'out of sight, out of mind' crisis. Groundwater is a major contributor to lakes and streams. As the water table drops lower so do lake and stream levels. Despite the warnings of the Department of Interior (DOI), and the US Geological Survey (USGS), "We continue to pump groundwater faster than it is replenished."

The USGS reports that "Groundwater is among the nation's most important natural resources. It provides half of our drinking water and is essential to the vitality of agriculture and industry. Large-scale development of ground-water resources with accompanying declines in water levels has led to concerns about the future availability of groundwater to meet domestic, agricultural, industrial, and environmental needs" (*USGS Ground-water availability in the United States 2008*).

Farming represents only 2% of the US population yet feeds the remaining 98%. About half the total population in the US and 90% of the rural population depends on groundwater. Farms and ranches consume over fifty billion gallons of fresh water per day. The amount of water necessary to produce a single food product is stunning. For example, it takes approximately two thousand gallons of water just to produce a single pound of beef. A recent study from the University of Twente in the Netherlands reported that, "The average water footprint per calorie of beef is twenty times larger than for cereals and starchy roots."

The ability to recharge groundwater is dependent upon precipitation patterns, but a changing climate is dramatically altering those patterns. In addition to population growth, the southwestern states are entering a mega-drought period that is projected to last for centuries into the future. As the Colorado and Lake Mead shrink, 40% of the nation's vegetables will disappear when southern California farms blow away. What happens to food production then?

In August 2013, scientists published their findings in the Proceedings of the National Academy of Sciences. They found that 72% of the major aquifers in Kansas will disappear by 2060. Once those aquifers are depleted the researchers estimated it would take anywhere from 500 to 1300 years to refill them and only then if they remain unused. Similar estimates were made for the states of Texas, Oklahoma, the Southwest, and the plains states. The impact on agriculture and cattle production is likely to be seriously threatened well before mid-century, meanwhile population growth will place ever greater demands on the water supplies. Freshwater shortages and pollution are two additional crises facing humanity by the end of this decade. In 2019, the World Economic Forum identified water scarcity as one of the greatest global risks by 2030. The connections and vulnerabilities between the demand for fresh water, food production, the economy, and security become clearer almost daily.

Mining and fracking practices by the fossil fuel industry are a significant threat to water shortages around the world. In the early 1970s, I watched Peabody Western Coal Mining Inc. take control of the last major groundwater source on Navajo and Hopi lands. Peabody Energy obtained exclusive access to the last groundwater on the Black Mesa plateau. Instead of using the water for agriculture or drinking, every year groundwater amounting to tens of thousands of acre-feet would be used to carry fourteen million tons of coal slurry in pipes to the coal-fired Mohave Generating Station 273 miles away. Now, some fifty years later, coal has lost its economic advantage, however, the Navajo aquifer has been severely compromised and the indigenous people who lived there for thousands of years find themselves in a mega-drought.

Water availability and distribution are only two factors affecting the importance of water. As the oceans warm more water vapor is evaporated into the atmosphere. For every 1.8°F (1°C) of warming, evaporation increases water vapor another 5% to 7%. The influence of increased water vapor on global and US precipitation is enormous and growing less favorable. Precipitation may now come with periods of extreme drought or sudden torrential floods.

Agricultural patterns of planting, growing, and harvesting are changing and difficult to predict or adapt to. Farmers try to compensate with drought resistant crops, increased irrigation, fertilizers, and other chemicals. Those measures may only increase the problem of pollution. Chemicals filter into water systems and reduce water quality. Some of the major food producing areas of the world are already threatened. The earliest civilizations faced this same problem. Continual intensive irrigation, over time, leaches the soil and poisons it with mineral salts. This happened in some of the first Mesopotamian civilizations over five thousand years ago.

The Sumer and Assyrian empires developed in the Fertile Crescent between the Tigris and Euphrates Rivers. The empires of bronze and iron rose and fell with the same pattern familiar to us today. For several millennia, the climate was moderate. These powerful civilizations grew and prospered. The bronze age Sumerians were among the first to master agriculture. They were displaced by iron and the military might of the Assyrians. Over the next five centuries the soil gradually became drier and less fertile. They adapted by developing irrigation on a massive scale. This provided another few centuries of prosperity until eventually over-irrigation and over-production leached mineral salts to the surface and sterilized the land.

Assyrian wealth and a powerful military allowed trade and conquest to continue to provide food but alienated their neighbors. The Assyrian solution was to crush, enslave, or exterminate their enemies. That approach backfired. Eventually, surrounded by hostile neighbors, powerful enemies arose in the Medians (later Persia) and the rebellious city-state of Babylon. The combination of vengeful neighbors, a changing climate, and the loss of soil fertility brought the beginning of their catastrophic downfall within a single generation.

In Southeast Asia today, rice growers are attempting to develop varieties that can be grown in brackish water caused by inundation from rising seas. Farmers plant more intensively only to further deplete the soil. The Mesopotamian scenario is being repeated around the world with similar results.

Even an extended growing season can be a problem as climate and local weather patterns shift. Water may come too soon, too late, all at once or not at all. Western US grain growers plant more land and develop varieties that can grow with less water. This same pattern led to the Dust Bowl of the 1930s, only to be repeated today with a sense of *déjà vu*. Water management in the US has been left primarily to the states. As water resources become threatened conflicts between local, state, and federal water resource management becomes more hostile and complicated. Colorado river water generates 40% of the power for Southern California and supplies water for agriculture and domestic water use from the Rockies to Mexico. As demand increases on this finite resource, the cost of water supplies will continue to skyrocket. Addressing anthropogenic influences on freshwater consumption and ocean pollution require national and international cooperation, but that too is threatened.

After the horrors of two world wars, the UN was formed and mandated to bring equity, justice, and peace through multi-lateral diplomacy for the common good. The harsh history lessons of Sumer, Rome, Athens, Egypt, and the Eastern Empires were all too clear. It was recognized that nationalistic policies of rule by force must give way to a philosophy of global governance through cooperation and mediation by all sovereign stakeholders. That has remained the UN mandate for more than seventy years.

In addition to my work as EPA liaison to the UN and World Bank Institute, I served for two years on the Board of Directors for the Seattle Chapter of the United Nations Association (now called the UN Foundation). I grew to appreciate the necessity, value, and complexity of international negotiations. During that time the UN was forming a language and goals for sustainable global development. Eventually seventeen interlinked, international sustainable development goals (SDGs) were ratified and published. Today they form the central unifying core of UN work around the world. In less than a decade from their inception, notable progress was made in every component of the SDGs. Then on September 24, 2019, President Trump addressed the 74th Session of the United Nations General Assembly.

US President Donald J. Trump stunned the world by advocating a giant leap backward toward nationalistic self-interest. Trump shunned the concept of a global community of nations and proclaimed a return to the same narrow territorialism that had contributed to both world wars and subsequent Cold War with the USSR. Mr. Trump declared: "The free world must embrace its national foundations. It must not attempt to erase them or replace them."

"Looking around and all over this large, magnificent planet, the truth is plain to see: If you want freedom, take pride in your country. If you want democracy, hold on to your sovereignty. And if you want peace, love your nation. Wise leaders always put the good of their own people and their own country first."

"The future does not belong to globalists. The future belongs to patriots. The future belongs to sovereign and independent nations who protect their citizens, respect their neighbors, and honor the differences that make each country special and unique."

I must confess experiencing a cold chill when I saw him posturing at the podium. Visions of a later-day Mussolini flashed into my mind. I heard an unpleasant and familiar echo from a divided and war-torn past. A past my father and millions of Americans risked their lives to put right. Many representatives of the of 197 nations apparently felt the same unease. Moments later, in secluded hallways and private chambers our G7 and NATO allies gathered and whispered in hushed tones, "I don't think we can count on America any longer" (German chancellor Angela Merkel).

Trump's UN declaration and US withdrawal from the Paris Climate Accord shocked the civilized world. It took place when virtually every nation on Earth recognized the vital necessity of addressing the existential threat of an overheating world. The leader of the most powerful nation on Earth, and greatest long-term contributor to global warming, had essentially said, 'abandon ship, its every nation for themself'. What the world heard between the lines of Trump's resurrected fascist double speak was, 'we've got ours, screw you'.

Trump's address was a blow to the very concept of shared interests and global coordination. He abandoned peaceful cooperation and development for an antiquated 'might makes right' foreign policy. The US government's own experts, and the agencies responsible for monitoring security, had an entirely different perspective. The great bio-geochemical systems of Earth don't recognize nationalistic bluster. No single nation owns the air above or the oceans that surround them. All nations must share the goods and services the environment provides. All nations must share the responsibility to sustain those goods and services for everyone's benefit, not the disproportionate benefit of a few or any one nation. Earth's oceans illustrate the importance of maintaining this key resource in common for all.

"The nation's valuable ocean ecosystems are being disrupted by increasing global temperatures through the loss of iconic and highly valued habitats and changes in species composition and food web structure. Ecosystem disruption will intensify as ocean warming, acidification, de-oxygenation, and other aspects of climate change increase. In the absence of significant reductions in carbon emissions, transformative impacts on ocean ecosystems cannot be avoided" (*2018, Fourth National Climate Assessment, Volume II; Impacts, Risks, and Adaptation in the United States*).

One of the most urgent UN sustainable development goals is to integrate and manage legitimate uses of global resources and protect those resources in a fair, equitable, and sustainable way. Those ideas are antithetical to Trumpian politics. Old enemies and some allies were beginning to see America's unregulated capitalism, authoritarian nationalism, and commodifying the commons as a threat to their own national and global security.

What if we returned to a world focused on national self-interest? What if major shipping lanes fell under the control of a few powerful nations? It happened before with devastating results. Great Britain, France, Spain, Portugal, Argentina, China, the US, and the Middle East are only a few examples where control of marine trade led to war. The goods and services provided by the Earth's waters are fundamental to the existence of all life.

Some nations view the global ocean as a vast untapped reservoir of resources for their own benefit. The seas provide a conduit connecting resources to the global economy. Today, over 90% of trade moves freely over the sea. The US may think it has shipping autonomy but most US marine trade moves in vessels owned by other nations. According to the Department of Transportation and the Maritime Security Program (Dec. 2017) the entire US civilian merchant fleet consisted of only 393 vessels, or twenty-seventh place in the world. Russia was in eleventh place with 1143 vessels, and China was second with 4052 merchant ships. The seas provide far more than conduits for trade. Perhaps America should not be preaching nationalism when it lacks the independent means of moving US goods to foreign consumers.

The Vikings, China's Zheng He, Magellan, Columbus, Captain Cook, and Alexander von Humboldt all gathered information on the biology, wind, and ocean current patterns during their great voyages of discovery. It is now a race between science and time to understand our influence on the marine environment before it passes several critical tipping points and feedback loops.

More than a third of the global population lives within sixty miles of the shore. Climate driven changes in the ocean alter the distribution, timing, and productivity of fishery species. The marine environment is a critical food source, providing roughly 15% of global food protein. Marine fisheries and communities that depend on fishing are at higher risk every year.

The ocean absorbs approximately 30% of human carbon emissions, and buffers global warming by absorbing over 90% of solar energy. The ocean currents moderate global temperatures. Ocean currents distribute nutrients where the great fisheries are found. This ocean conveyor system determines wind and precipitation patterns, and the climate over the Earth's continents. Without these services the planet would be uninhabitable. It is precisely those services that are now endangered.

Every other breath you take comes from the ocean. Ocean acidification, pollution, and antiquated agricultural management practices are now beginning to threaten the life cycles of oxygen producing marine plant life. Most people don't realize that marine plants produce more oxygen than all the green plants, forests, and jungles on land combined. If acidification and pollution continue to increase, it is inevitable that sometime in the future marine oxygen producers will experience intolerable stress.

Acidification will eventually lower metabolic efficiency in photosynthetic plankton, reducing phytoplankton's ability to metabolize carbon dioxide and produce oxygen (Crawley and Kline, 2010; Brading and Warner, 2012). The question is, will oxygen depletion be sudden or gradual? Brian Palmer, science reporter for the National Resource Defense Counsel puts it this way: "...do you like oxygen?"

As the Arctic Ocean grows more ice free, nations vie for control of newly exposed and unregulated resources and shipping lanes. In southern seas, pirates still plunder. Despite the ocean's vital importance, there are very few international maritime laws to ensure the ocean is adequately protected. Enforcement is almost entirely voluntary. The ocean and atmosphere form an international common that must be protected collectively by all nations.

Few laws effectively protect marine species, though nearly all the large marine species like swordfish, salmon, Atlantic bluefin tuna, Atlantic cod, Atlantic halibut, European eel, European sturgeon, grouper, haddock, northern red snapper, Giant Sea Bass, and 149 other species are endangered or threatened (NOAA). Whaling bans serve only as guidelines for most nations and are totally ignored under false pretenses by others. Enforcement is often left up to private, non-profit organizations like Sea Shepherd, Green Peace, World Wildlife Fund (WWF), and other Non-Government Organizations (NGOs). When private non-profits step in, there is a narrow path, they must tread between hero and villain, even for the most noble of causes.

The Grand Banks cod fishery is a case study in over-fishing. In the 19th and early 20th centuries, cod was a staple in the American diet. North Atlantic cod fishing had flourished for nearly five hundred years. Fisheries like the

Grand Banks were considered inexhaustible. Codfish oil was considered a miracle cure and source of vitality and long life.

I first read *Captains Courageous* when I was five. Kipling gave an accurate account of Gloucester fishermen, their almost mythical schooners, and the seemingly endless bounty of the Georgian and Grand Banks cod fishery. Men in sleek wooden dories baited hooks and pulled hand lines until the dories were awash with flapping tails. The dories fed the schooners where the fish were processed, salted, and put into the hold.

These sleek schooners, sporting sails measuring thousands of square feet, were euphemistically called 'knockabouts'. They were deceptively fast, really fast. Once filled to the gunwales with cod, they raced to market. Those that arrived at port first could ask the highest price.

Knockabouts were so quick, that in the offseason, they took on the best that international yacht racing could offer. Perhaps the most famous of these working-class schooners was the 'Bluenose'. For two decades, no other design could compete. One International yachtsman was asked why he lost to the 'Bluenose'. He replied, "We race for fun. These fishermen do it for a living."

A single 'knockabout' could salt up to 250,000 pounds of cod in their hold. The work was hard and dangerous. Each year as many as four hundred 'knockabouts' went to sea. Each year a dozen or more of these sleek craft never returned.

Fishing vessels and equipment made a great leap forward when power replaced sail. More vessels and better equipment brought in more fish. Gradually a fishery decline became apparent. By 1970, foreign fishing vessels were banned, but American fishing continued the harvest with even more efficient technologies. What was thought to be inexhaustible, wasn't. By the 1990s, both Canada and the US agreed to close fishing for all bottom-dwelling species, including cod. There were hopes that the fishery would recover but recovery did not happen. Warming waters in the Atlantic further compromised the resilience of cod stocks.

Once an ecosystem adjusts to new conditions, it can never return to what it was before. Other species soon replaced the cod niche in the ecosystem. The cards had been shuffled and a new game had begun. Yet, the demand for cod remained high with increasing pressures on stocks elsewhere. Four factors appear to be primarily responsible for a fishery collapse:

1. Improvements in fishing technology.
2. Uncertainty of assessing fishery resources vulnerability.
3. Socioeconomic factors (demand) that conflict with fishery management.
4. Lack of Government oversight and regulatory authority.

The history of cod and the Grand Banks fishery serves as a lesson for fisheries in general. In the absence of international protections for the ocean commons, some nations are expanding their territorial waters to protect vital fishery resources and trade routes. It remains to be seen how long those remaining fisheries will last under an ever-increasing demand.

Nationalism has increased the threat to international marine security. When President Trump claimed nations should set self-interest as their highest priority, it opened a nationalistic free-for-all competition for marine resources and control. The Spratly Islands southeast of China are over five hundred miles from the mainland. China will use these islands to justify control of new areas of the South China Sea. The speed and scale of this move has alarmed the international community and surrounding nations of Vietnam, the Philippines, Malaysia, and Japan. Satellite surveillance shows the construction of airstrips, port facilities, and military assets. In less than a decade, China turned these distant shallow seamounts into militarized bases and vastly expanded their territorial influence. This threatens access to productive fisheries and free passage for international trade through one of the most active shipping lanes. It is a breach of international maritime law, but who is willing to enforce the law against a powerful nation like China?

Other nations have used more covert hit and run tactics to supplement diminishing fish stocks. Fishing vessels must travel farther to find their prey. Frequently the fishery rests in someone else's territorial waters. Foreign fishing vessels have routinely invaded US waters for more than two centuries. If pursued, they drop their nets and skedaddle. The abandoned nets continue to kill fish for decades, further depleting fish stocks. These ghost nets pose a hazard to marine navigation and make up approximately 10% of all marine litter.

In many parts of the world, fishing is already unsustainable because of a lack of oversight, regulatory control, and sound science-based policy. Economic pressure and ever-increasing consumption are core problems in fishery management. At least, a quarter of all fish stocks are overexploited or dangerously depleted. Nearly 80% of larger fish species are already endangered. Illegal practices and overfishing continue.

As fishing techniques become more efficient, they continue to catch other species (bycatch) that are simply treated as waste despite their importance to marine ecology. (Bycatch refers to other marine species that are caught unintentionally. Bycatch is either of a different species, the wrong sex, undersized, or juveniles of the target species.)

Overfishing, ocean warming, and acidification are all throwing marine ecology seriously out of balance. Acidification and warming temperatures are a death-knell to marine coral. Coral reefs are a critical habitat for 25% of marine life. Recently ocean temperatures reached 101.1°F (38.9°C) offshore near Maimi, Florida. Bottom dredging and mining coral for cement production destroys habitat that may not recover for years, if ever.

Eliminating greenhouse gas emissions and installing proper regulatory oversight can save most fisheries if it is accomplished before ocean acidification, Arctic warming, and deoxygenation permanently destroy the marine food chain. Without a thorough understanding of system linkages, regulatory oversight, and strong international enforcement, the problems of pollution, bycatch, and mismanagement will continue to threaten production of our entire marine ecology and global fishing industry.

In the movie The Graduate, Mr. McGuire said, "There is a great future in plastics. Think about it. Will you think about it?" Apparently, we didn't think about it enough. In addition to the threat of overfishing and ocean acidification, roughly sixteen billion pounds of plastics were dumped into the ocean in 2020. This was in addition to the one hundred fifty million tons already contributed. The total tonnage per year is expected to triple by 2040. Most plastics will eventually crumble into micro-particulates that pose another hazard. Microscopic plastics are now beginning to appear in fish tissue that ends up on our dinner plates.

Dead zones are hypoxic ocean patches devoid of oxygen where few things can live. They are found around the world, but more importantly to the United States, they are now occurring off the coastline from Washington State to the Gulf of Mexico and up the eastern seaboard. As the ocean warms it loses its capacity to hold oxygen. The combination of warming ocean, acidification, and runoff is causing the ocean to run out of oxygen. This has a direct impact on marine fisheries because fish need oxygen just like we do. The largest dead zone in the world covers 63,700 square miles of the Gulf of Oman in the Arabian Sea. The largest US west coast dead zone encompasses over 27,000 square miles.

According to Monica Bruckner, Project Specialist at the Science Education Resource Center (SERC) at Carleton College, nitrogen and phosphorous runoff used in agriculture contribute to dead zones. One of the worst examples of chemical leeching and runoff is the Mississippi watershed. The Mississippi and its tributaries drain the central United States from the Canadian border to the Gulf of Mexico. Despite this, agriculture must keep up with food demands to feed a growing population.

Fertilizer runoff provides nutrients that allow plankton to bloom in such large numbers they influence the entire ecology of the Gulf of Mexico. Once plankton has consumed all the available nutrients, there is a massive die-off that settles to the bottom to decompose. Decomposition consumes the remaining dissolved oxygen in the water producing a dead zone. According to NOAA, in 2017 the dead zone near the mouth of the Mississippi covered more than 8,000 square miles (about the size of New Jersey) and cost billions of dollars in lost fishery revenue. In the last fifty years, the number of dead zones has quadrupled from forty-nine in the 1960s to more than four hundred in 2019. In recent decades, climate change has led to an unprecedented increase in flooding throughout the entire Mississippi watershed. NOAA forecasts indicate flooding will continue to set records in frequency and volume. The increasing use of agricultural chemicals, combined with record flooding and poor regulatory control of runoff, has increased the nitrogen content of the Mississippi to three times what it was in the 1950s. Phosphorous concentrations have doubled over the same period.

Warming oceans pose another problem. Warm water can't hold as much oxygen as colder waters. As warm water loses the ability to hold oxygen, it disrupts the ecological balance between animal and plant plankton. This affects larger organisms directly, but also threatens the smallest marine life. Plankton and Krill form the base of the food pyramid. Many marine organisms graze on phytoplankton (plant plankton) just as herbivores graze on the land.

Any disturbance in the stability at the base of the food pyramid impacts everything above it from sardine to salmon to whales to humans. As lower latitudes warm water forces plankton and fish to migrate to cooler more oxygen rich waters.

The ocean absorbs a third of human carbon emissions. That carbon converts to carbonic acid. Ocean acidification has risen nearly 30% from pre-industrial levels and is beginning to prevent the formation of shells and exoskeletons in krill, mollusks, crabs, corals, and many plankton species. This change is occurring more rapidly than many species can adapt to, further contributing to the ongoing sixth mass extinction.

As the planet warms there are other factors that contribute to ocean acidification. As the Arctic warms and glaciers melt, it changes both horizontal and vertical ocean circulation. Some of these currents stir up accumulated carbon sediments from animal and plankton remains that settled to the ocean floor over millions of years. This added carbon further contributes to the acidification of marine waters. Model projections indicate that ocean acidification could increase as much as 150% by the end of the century.

The problem of increased demand for fish protein is taking place while the marine environment is under severe stress from warming, acidification, pollution, and expanding dead zones. This contribution to mass extinction includes nearly all the species we like to eat.

Replacing wild production with aquaculture creates a host of other management problems. Hatchery-raised species may not be as nutritious as wild species. They tend to be less robust and more prone to disease. Ask any trout fisherman if hatchery fish are as scrappy as wild fish. Fish farmers must feed the fish, but where does the feed come from? What do we do with fish waste?

Washington state's seafood industry generates more than 43,000 jobs and $1.7 billion annually. Pacific Northwest Shellfish farmers have suffered up to an 80% loss in production because of marine acidification. Greg Dale, southwest operations manager at Coast Seafoods Company in Northern California summed it up this way: "The ocean drives everything—the whole food web, our weather, clouds."

The Earth's oceans have been used as a toilet and garbage dump since the beginning of civilization. The bill for that negligence has come due. What once seemed vast and imperturbable is now clearly vulnerable to increasingly toxic shortsightedness and the flotsam and jetsam of self-interest nationalism.

The ocean drives everything. What happens in the ocean has a direct impact on the climate, the food we eat, the water we drink, and over half of the oxygen we breathe. The rapid increase of carbon dioxide emissions over the last fifty years poses a major threat to life in the sea and has the potential to literally suffocate life on land. The story of cod fishing, warming seas, acidification, and hypoxic zones are just a few examples of what is happening with the entire marine ecosystem. The once inexhaustible ocean commons has proven vulnerable to special-interest exploitation.

We must choose between cooperative transnational stewardship or to be overwhelmed by environmental collapse. To solve the looming crises of this relationship we must first understand those linkages. To do that, we must become system thinkers.

Chapter 7
System Thinking

Most of us have a hard time seeing the complex linkages between bio-geochemical planetary systems and our daily lives. That doesn't exempt us from the necessity for humanity to recognize the interdependence between civilization and the environment. We are now faced with the necessity of growing more food, on less land, with less water, and a climate that is no longer hospitable. We need to be informed.

If you ask John or Jane Doe where milk comes from, they might say cows, but the correct answer is far more complex. The sun provides 100% of the energy that drives every bio-geochemical system on earth. These planetary systems provide the precipitation, winds, and erosion that release the minerals that feed the micro-organisms in the soil, that release the nutrients, that grow the grass, that feeds the cow, that produces the milk that we buy in the store. Even this description is an oversimplification. The linkages are infinite.

In the 1960s, the world was faced with a serious food production crisis. American agronomist Norman Borlaug began developing new techniques of plant breeding that allowed much higher production on the same area of land. For this, he received a Nobel Peace Prize in 1970. Norman Borlaug's Green Revolution was only a temporary solution.

Civilization is at the center of a vast multi-dimensional, planetary web. The human economy plucks strand after strand, not knowing when it will reach that one strand that causes the entire web to collapse. The footprint of more than eight billion people is at once obvious and bewildering in its complexity. We act before we fully understand the systems involved or the consequences. Policies are often knee-jerk, short term, or too narrow in scope because policy makers failed to understand the need for system thinking.

Earth's dynamic equilibrium and resilience to human insult is near collapse. Earth is a garden that civilization must attend, not a grocery store.

It is an axiom of nature that all species, including humans, reproduce until they reach the limits of available resources. As our population has grown, food consumption has grown even faster. Prosperity has moved us higher up the food pyramid. We eat more sugar, fats, and meat than ever before. To keep up with demand, we clear more land, add more chemicals, and genetically modify food crops. These practices compromise the natural systems that replenish the soils fertility. Cities, highways, and suburbs cover prime agricultural land. Climate change increases desertification, further reducing the amount of land available for food production. As a result, the total amount of arable land is decreasing.

It makes little sense to pour more chemicals onto the land when uncontrolled runoff threatens the rivers, lakes, and seas around us. The long-term consequences are overwhelming. Some of the finite minerals necessary for current agricultural practices are running low. Farmers around the world are already experiencing shortages and higher costs. It becomes an environmental justice issue when prices rise beyond the purchasing power of major portions of the global population.

One of these finite minerals is Phospate. Phosphorous is necessary to produce chemical fertilizers. In 1961, phosphate went for sixty dollars a ton. In 2015, the price was seven hundred dollars. Phospate is abundant but it is increasingly difficult to mine at low cost. With a growing population and demands to increase food production, accessibility of rock phosphate reserves and rising cost is a serious concern. Nearly all phosphate is recycled by nature but current agricultural practices interrupt that cycle. System thinking tells us that it is necessary for another major revolution in agriculture.

Nature has been renewing soils for eons. Indigenous people learned to use the natural ecology of soil to prevent nutrient runoff and restore fertility. Agricultural science and farmers recognize they are in a race to develop these techniques in time to keep up with population growth and declining agricultural mineral reserves. We are already well into the sixth and most rapid mass extinction in Earth's history. This is because of environmental disruption, over consumption and population growth. We are faced with the consequences of that reality. How bad is it?

The Permian extinction or The Great Dying occurred approximately two hundred fifty million years ago. It wiped out 95% of all life on Earth. As temperatures rose warmer waters could not hold enough oxygen for marine organisms to survive. (Dec. 7, 2018 issue of Science). Part of the cause for the Great Dying was a loss of marine plants that reduced the amount of oxygen available to the Earth's oceans and atmosphere. The alarming fact is that the present rate of extinction is over a hundred times faster than any previous extinction event. Give that a think for a second. How could humanity survive if 95% of life were to go extinct, especially if it happened in a matter of decades instead of millennia?

Since *Homo sapiens* spread over the entire planet, 55% of all previously known large animal species became extinct. It has only taken the last two hundred years to drive another 50% to extinction. Precise causation for those extinctions is difficult, but the correlation between extinction and the rise of humanity is clear. Biologists now predict that one million species are likely to become extinct this century. It is now probable, in this generation, that there will no longer be wild elephants roaming freely over the African savanna. There will no longer be eighteen foot tall giraffes or wild African lions or Siberian tigers. We will no longer see White Rhinos crashing, like tanks, through thorny brush to reach a nearby watering hole. Our children will only see them in zoos or pictures. Carolina Parakeets, Passenger Pigeons, Dodos, Elephant Birds, Steller's Sea Cows and the Western Black Rhinoceros are all gone. In their case, human causation was clear.

An ocean in peril and the sixth mass extinction are only part of civilization's footprint on the four realms of Earth. The air is more polluted, and the frozen parts are thawing. The polar caps serve as Earth's air conditioners. They keep the atmosphere and ocean from overheating. With less ice, there is less solar energy reflected to space (albedo). More energy is absorbed by the darker water causing the ocean to warm. Warming then causes marine currents to alter their depth and circulation patterns. As the ocean currents change so does the climate.

What happens in the Arctic does not stay in the Arctic. The Arctic is warming an average of four times faster than lower latitudes, with some regions warming up to twelve times faster. Coastal glaciers and sea ice are melting from heat energy applied from above and below. The Arctic ice cap is not only smaller in area each summer, but also thinner, with less holdover ice from previous years. The reduction in total ice volume has a worldwide impact.

The warm waters of the Atlantic Meridional Overturning Circulation (AMOC) currents are responsible for the moderate climate of Great Britain,

France, Spain, and the Nordic nations. The AMOC is responsible for bringing warm water from the Gulf of Mexico and warmer latitudes to the west coast of Europe. Fresh water melt from Greenland and the loss of polar ice appears to be altering the strength and circulation pattern of the AMOC. Stefan Rahmstorf, lead Physics of the Oceans Professor at Potsdam University in Germany believes the AMOC may be at or past the point where it begins to shut down.

Greenland is warming much faster than anticipated. Sea ice at the foot of land-based glaciers serve as a dam that slows movement seaward. Warmer ocean water weakens the dam and allows the glacier to flow faster. In some cases, water may move inland under coastal glaciers lubricating their slide to the sea. Warm air above melts the glacier's surface, forming melt ponds called moulins. As these moulins are warmed by the sun, they cut vertical shafts leading under the glacier. All these factors are occurring in Greenland at rates much faster than predicted only three or four years ago.

In the spring of 2020, temperatures above the Arctic Circle averaged 14.9°F (8.3°C) above the thirty-year norm. Temperatures over 100°F have been recorded north of the Arctic Circle in Siberia. Argentina's Esperanza research station on Antarctica's Trinity peninsula recently recorded T-shirt temperatures as high as 64.9°F (18.3°C). To get an idea what this does to the ecology in these regions, imagine the summer temperatures where you live was 14.9°F (8.3°C) warmer than your previous record for a really hot summer day. I live near Seattle, Washington. Last year the heat record was set at 108°F (42.2°C). Now add 14.9°F (8.3°C) to that. A hot summer day in Seattle would then be 122.9°F (50.5°C). If that were to continue, Washington State would no longer be the land of green forests and cascades. It would become more like Death Valley.

The Thwaites glacier, sometimes referred to as the Doomsday Glacier, on Antarctica's Southwest coast, recently grabbed the attention of scientists. The glacier's rate of movement dramatically increased. The sea ice shelf, in front of the Thwaites glacier, began disintegrating. Masses of ice the size of Delaware state began to break off. Ice bore holes and ice penetrating radar revealed that the land under the glacier was retrograde (sloping inland instead of seaward). As the ice shelf weakened, seawater began to flow inland under the glacier, further lubricating and accelerating the glacier's slide into the sea. To make matters worse, the water seeping under the glacier started to erode an enormous chamber, raising fear that the glacier's own weight might cause the roof of the chamber to collapse.

The catchment basin holding the Thwaites glacier is about the size of Great Britain. It serves to block the far larger West Antarctic Ice Sheet. The collapse of the West Antarctic Ice Sheet was projected to be centuries, if not thousands of years off. Without the Thwaites glacier to hold it back, collapse could begin far more quickly.

The rate of land ice movement is increasing. Total collapse of the Thwaites glacier would raise global sea levels by as much as three feet (0.9 meters) inundating millions of square miles of low-lying coastal lands around the world. High tides and storm surges would impact hundreds of millions of people farther inland. *Au revoir* New Orleans, *hasta la vista* Miami, and hundreds of other coastal cities around the world.

This is a problem researchers and modelers are grappling with. A tipping point may have already been reached where further melting and subsequent sea level rise is inevitable. Just when we think we have a handle on things, another strand in the bio-geochemical web snaps and we must recalibrate. Once again, the question is when, not if.

The good news is that rising sea levels do not happen overnight. That should not make us complacent. If we act now, coastal cities should have time to migrate to higher ground. Any further procrastination, however, would turn a stroll into a sprint.

The Arctic is an entirely different and far more pressing issue. The time for an easy transition has already passed. The Arctic is warming faster than Antarctica. This rapid acceleration is called Arctic amplification. Catastrophic regional and planetary changes are already happening. Arctic amplification is a cause for urgent concern while it is still possible to mitigate some of the impact. A warming Arctic causes huge changes in atmospheric and ocean circulation. For example, the atmospheric temperature differences between the Arctic and temperate latitudes were quite different, until recently. That difference tended to form a stable atmospheric pressure barrier. This phenomenon is like the thermocline you experience wading along the shore of a lake and your feet suddenly feel a layer of colder water. The atmospheric zone of high temperature and pressure differences created a fast moving, high-altitude river of air. These fast-moving currents are called the jet stream.

Arctic amplification appears to be destabilizing the jet stream and causing it to undulate, contributing to another phenomenon called the polar vortex. The current hypothesis suggests that the polar vortex may be the result of less temperature difference between Arctic and temperate air masses.

The weaker barrier allows wildly fluctuating undulations of unseasonably warm or cold air to push farther north and south, hovering for weeks at a time. That undulation is believed to be the cause of recent record cold temperatures from central Canada throughout the Midwest and Eastern Seaboard of the US. Conversely warm air masses may also bulge into Arctic regions. Similar events are taking place in Europe and Asia.

In late June of 2021, a heat dome hovered over the US shattering all-time records on both coasts. On the west coast, Seattle had record triple digit temperatures for nearly a week. Six months earlier Seattle temperatures in December were the coldest in thirty-one years. Portland, Oregon had temperatures reaching 115°F (46.1°C). As anthropogenic warming increases, these extreme events will become more widespread, powerful, and frequent, threatening ecosystems, infrastructure, and the economy.

As temperatures rise, land locked glaciers pose a threat of a different kind. Glacial ice in the mountainous regions of our planet form the headwaters of most of the world's great rivers. The glaciers in the Rocky Mountains give birth to the Missouri, Mississippi, Colorado, and the Columbia. The Alps, Urals, and Carpathians provide water to most of Europe. The Andean glaciers give rise to the Amazon and its tributaries. These rivers are vital to food production for more than a billion people, but they pale next to the Himalayas. Nineteen major rivers drain from these peaks. The largest are the Brahmaputra, Indus, Ganga (Ganges), Irrawaddy, Mekong, Yangtze, and Salween. They determine the water, food, and energy production for nearly three billion people. As the Himalayan glaciers melt, China, India, Pakistan, Thailand, Vietnam, Laos, Myanmar, Bhutan, and Cambodia begin to go thirsty and hungry. It is important to note that the first three nations are nuclear powers with a history of animosity toward each other.

Migratory birds and animals depend upon the climate and predictable seasons for migration, food, and reproduction. Caribou herds need solid ground to migrate. Arctic amplification is melting the permafrost and turning it into bogs that block migration routes that existed for thousands of years. Alternative migration routes are blocked by fences, pipelines, roads, and settlements. Waterfowl need favorable conditions for nesting. In a warmer climate, some species are choosing to winter over in Canada and the US instead of migrating to southern states or Mexico. This can turn waterfowl migration into an agricultural and urban 'pest' problem.

Permafrost is melting much faster than predicted less than a decade ago. Then it was merely interesting. Now the threat of an unstoppable feedback loop has grabbed the attention of scientists. When permafrost melts, it begins to decompose, releasing carbon dioxide and methane. The question is how fast and how much methane is released. Methane has eighty-four times the CO_2 GHG equivalency.

Methane doesn't linger in the atmosphere like carbon dioxide. Within a decade, most methane has broken down. This brings the rate of warming into the spotlight. To become a serious danger, abrupt methane emissions would have to occur within a few decades or less. For half a century, scientists thought that a rapid release of methane from permafrost was unlikely. Now some are beginning to believe it is a definite probability and may already be underway.

Unprecedented Arctic warming in 2021 and 2022 began to release massive amounts of permafrost methane. In Siberia bubbles (domes like gigantic pimples) began to appear under the permafrost. When these methane pimples burst, they left enormous pock marks over the landscape. Some were more than 150 feet deep and 100 feet wide. As this phenomenon continues, it could lead to a feedback loop of more methane, causing more warming, releasing more methane, etc. Once that feedback loop is underway, methane and carbon dioxide releases from melting permafrost could take over as the dominant driver of Arctic warming. Whatever civilization did after that to stop anthropogenic CO_2 emissions would have little effect.

This chapter described the complexity of Earth systems and how the interactions between those systems are difficult to explain to the general population. Our normal perceptions are so limited, it's difficult to see the whole problem. There is a parable that originated on the Indian subcontinent. It's about three blind men who try to describe an elephant by touch alone. Each man feels only one part of the elephant. One man feels the elephant's side. Another feels a tusk. The last man feels the elephant's leg. They each describe the elephant based upon their limited experience. Their descriptions are sincere but entirely different. The men begin to suspect that the others are dishonest and they come to blows.

The moral is that humans tend to claim absolute truth based upon limited and subjective experience. They ignore other limited, subjective experiences that may be just as true but just as incomplete. Our senses and limited experiences may be sincere but still mislead us. For example, a sunny Nebraska day in May gives little clue that it is the beginning of winter south of the equator. The scorching January fires in an Australia summer are difficult for a shivering Fargo North Dakotan to understand or sympathize with. Most people have a vague idea that the ecological footprint of eight billion people must be enormous. As urban centers expand more potentially arable land is lost. Cities, farms, and industry drain rivers, lakes, and ground water faster than it is replenished. Forests and wildlife habitats are ravaged. Migration routes are blocked with fences and roads. Wild game becomes scarcer every year. We don't personally experience all of these, so like the blind man who only touches part of the elephant, the larger picture eludes us.

Devastating fires, floods, hurricanes, and droughts sound so extreme, many of us suspect they are exaggerations, even when they are accurately reported. The planetary systems that replenish the land, protect our water, and renew the forests don't fully register in our civilization insulated lives. We hear of extremes but unless one happens to us, it doesn't sink in as being real.

The most current models indicate that warm areas of the globe should continue to become warmer and wet areas should continue to become wetter. But a changing climate is not uniform because the Earth's surface is not uniform. The rotation and tilt of Earth's axis influence heat distribution and climate patterns. We commonly recognize these zones as the tropics, the dry latitudes where we find the great deserts, and the temperate and polar latitudes. These bands are expanding pole-ward as the ocean absorbs more heat energy. Deserts expand into regions that were previously savanna and forest (desertification). Temperate forest regions expand into areas that were previously frozen tundra. Lands that were previously frozen in permafrost are turning into methane and carbon dioxide generating bogs. Changes in monsoon and storm patterns are threatening food security for nations around the world. The effect of these changes has a potent destabilizing linkage to social, environmental, and economic security.

The linkages between energy and security are now obvious to NATO nations, the EU and most of the rest of the world. In the 1970s and 80s, we thought we could make an easy transition from fossil fuels to a cleaner more sustainable civilization. That didn't go so well. Now we must deal with the Ukraine war where fossil fuels are once again threatening global security. Putin thought Russia would get away with it because of EU dependence on Russian natural gas.

"As rising global temperatures change the monsoon and cyclone patterns in South Asia, the impact on economic growth will only worsen. Populations in climate-vulnerable areas are stuck. Before they can recover from a disaster caused by one natural hazard, another one strikes, reversing any progress that was made. To end this cycle, South Asian communities need to build greater resilience. As they confront the exorbitant cost of disasters, South Asian governments are beginning to realize the benefits of resilience. But to do that, the region needs to adopt ambitious policies and strengthen planning" (The World Bank).

For decades, I had an academic understanding of climate change, but it didn't fully register with me until I traveled enough to experience it personally. I've now witnessed the impact of a changing climate in China, Canada, Nepal, India, the Middle East, North Africa, and nations within the Himalayan watershed.

My adventure travel business partner for more than thirty years was Lopsang Tsering Sherpa. Lopsang is a nephew of the late Tensing Norgay, who along with Sir Edmond Hillary, were the first to summit Mt. Everest in 1953. Lopsang and I were friends with Narwang Gombu, another great Sherpa expeditioner and Director of the Himalayan Mountaineering Institute in Darjeeling. On the trail Gombu, Lopsang and I often sat around the campfire sharing stories and drinking yak butter tea. Back in Kathmandu or Darjeeling we would pour over more than eighty years of photographs showing glaciers and the treacherous approaches to many notable and notorious Himalayan summits. Those photographs showed decades of shrinking glaciers and an alarming change in ecology throughout the Himalayas.

The rate of glacial melting has doubled in the past twenty years. Josh Maurer of Columbia University's Lamont-Doherty Earth Observatory authored a study that looked at images of 650 individual glaciers and found on average they had been losing a foot and a half in depth every year from 2000 to 2022. Some glaciers had shrunk in depth by hundreds of feet. A similar study by more than 250 researchers predicted that the Himalayas could lose two-thirds of their mass by the end of the century. Water shortages already threaten food production and potable water supplies in Southeast Asian nations and the Indian subcontinent.

There is the false assumption that wealthier nations will have a much greater resilience and little difficulty adapting to any crisis. The COVID pandemic showed that even the richest nations can be brought to their knees by a single event. After three years of COVID, supply chains are still shattered. Water, soil depletion, crop failures, droughts, storms, and the constant threat of new pandemics assault the world every day. They are all magnified by climate change. America's failure to address COVID may be the best proof that not even the richest and most powerful can stand alone against such threats.

If the deaths of over one million Americans and six and a half million around the world isn't alarming, it's because civilization insulates most of the public from the sharp edges of reality. America's longest war in Iraq and Afghanistan was horrendous, yet the average US citizen continued with life as usual because less than 1% of the population had any connection to war. Nearly all of the financial cost of the war was passed to future generations.

In the 1950s, Smallpox, Polio, and Measles were identified as a public health crisis. The public was informed from a wide variety of sources. The drive for vaccinations was broadly publicized as a national public responsibility. Most people accepted vaccinations as common sense and a patriotic duty. The result was the total eradication of smallpox. The public acceptance of vaccinations brought the pandemic threats of polio and measles under control.

Between the year 2000 and the present, numerous studies revealed that the relative security of civilization tended to spread public apathy. As the feeling of security grew, indifference grew in proportion. America today is quite different from America in the 1950s. The sense of community and national responsibility has changed. The 1980s became the 'me' generation. By the 21st century, individual rights began to supersede everyone else's. Patriotism became individual rights and not the rights and wellbeing of community. The refusal to take socially responsible precautions against COVID was an example. By refusing to get vaccinated, a small minority placed the majority at higher risk. When people believe their individual rights and beliefs are more important than everyone else's, no one is free. Think of a boat with a dozen people on board. There are also a dozen leaks. If eleven people stop eleven leaks, the boat will still sink because one person chose not to stop their leak.

You can't get much more personal than attacking the autonomy of another person's body. In 2022, fifty years of reproductive rights were taken away when Row v. Wade was rescinded by a six to three conservative Supreme Court. Roughly 30% of the decision making population (mostly men) believed their values should govern the rights of the remaining 70% (mostly women).

Today a minority group of special interest corporations and a handful of the extremely wealthy threaten civilization itself and inhibit responsible international action to mitigate the most pressing crises of our time. Oligarchs prefer authoritarian leaders because it is easier to negotiate reciprocity between their interests. We are in a battle where the special interests for a few threaten the survival of the majority. Just as a small minority enforced their personal reproductive choice on all women, an economically influential minority is forcing the world to experience the life-threatening crisis of climate change, mass extinction, economic and political insecurity.

Moderately dry areas like the Mediterranean countries of Spain, Italy, Greece, and North Africa are already experiencing unprecedented heat waves and droughts. Under the most optimistic scenarios, parts of the Middle East are projected to become seasonally uninhabitable in the latter half of this century. It is very likely to happen much earlier if emissions remain on a BAU pathway.

Global economies are so intertwined that heat stress already compromises the security, health, and economies of all nations. Tens of millions already flee parts of the Middle East and North Africa. In thirty years, parts of Italy, France, Spain, North Africa and the Middle East are likely to experience nearly intolerable heat stress. What happens to the people who live there? Can they accept more refugees, or will they become refugees themselves? Can we allow a powerful economic minority (fossil fuel lobby) to determine the health, autonomy and wellbeing of the rest of the world?

Chapter 8
The Decadence Gene

When I taught science for the Bureau of Indian Affairs, I obtained a Title-9 grant to take twenty Navajo students to San Francisco for a lecture on man in the environment. One of the presenters droned on about *Homo sapiens* eventually rising from primitive hunter-gatherers to the pinnacle of civilization (emphasis European). The presenter then turned to American history. He believed European influence was entirely responsible for advancements in the Americas. At this point, one of my students slowly stood up and raised his hand. The speaker stopped and looked out over the packed audience.

"Yes…you have a question?" the presenter asked.

"You said primitive humans and the great apes had a common ancestor."

"Well, yes, apes and primitive humans even defecated in their nests and caves. I'd say marking your territory with feces could be considered primitive behavior," he replied with a somewhat sarcastic tone.

"You said apes evolved in one direction and humans went in another to become hunter-gatherers like the indigenous peoples of the Americas. You then said that another split occurred where the hunter-gatherers were sort of frozen in time while more advanced humans went on to develop civilizations like those in Europe."

"Yes."

"You also seemed to suggest that the indigenous people of the Americas, like me, weren't sophisticated or as advanced as Europeans of that time."

"Well, there were some developments…" he started to say.

Interrupting, my student continued. "The Americas weren't discovered by Europeans. My people have been here for over twenty thousand years. There were roughly sixty million indigenous people thriving here when Cortez arrived at the Aztec capital of Tenochtitlan (now Mexico City). The population of that city was between 200,000 and 400,000, making it one of the largest cities in the world. It was larger, richer, and more scientifically sophisticated than Madrid or London. People in London were still living with their livestock. European priorities were conquest and riches, not improving civilization. It was European weapons and disease that destroyed my ancestors not superior civilization."

"By the nineteenth century, European diseases had reduced the indigenous population of the Americas by 90%. In my short lifetime, I've studied how so-called civilized Europeans destroyed the American environment in a few centuries. In the last century, European civilization dragged the entire world into war, twice. Many indigenous cultures in the Americas were as wise as any European. The US Bill of Rights was copied from the thirteen confederated tribes of the Iroquois."

"Our trade routes covered the entire Western Hemisphere from South America to Alaska. Our land remained rich and fertile, and the game plentiful because we believed we should live as part of the natural world. The modern, cultured, sophisticated civilization that you seem to praise, robbed indigenous peoples around the world, polluted the land, cut down the forests, killed the animals and continue to waste, pollute, and crap in their rivers and drinking water. Are you sure the civilization you speak of is the best example of what a thriving and sustainable civilization should be?"

My class stood up, quickly followed by the entire audience, who clapped and cheered.

That was more than fifty years ago. My student's point raised the question that is the thesis of this book. There are millions of young people in the streets, raising their fists and crying, "This is all wrong," while the rich and powerful lie to them, call them naïve, and steal their future. Are we sure western civilization is as civilized as we think it is? Has civilization caused us to lose our sense of community and with that our sense of empathy? Arnold Toynbee said, "Great civilizations are not murdered. Instead, they take their own lives."

Writer Sebastian Junger explained why he thought returning American soldiers had a suicide rate twenty-five times higher than Israeli soldiers. "Basically, soldiers in combat experience something that's a close reproduction of our evolutionary past. We evolved to live in kinship groups of thirty, forty, or fifty people functioning very closely, sleeping together, eating together, doing everything together. Our survival depended on the group. That's our evolutionary past. It's also life in combat. It's even life in a platoon at a rear base. Most of the military doesn't fire their weapons at the enemy, don't get shot, but they do function in these close, tight-knit groups, and those emotional bonds become incredibly important. That's what we're wired for." Those bonds are broken when Americans soldiers return to the much looser connections and anonymity of US civilization.

Part of the definition of civilization is specialization. People in the 21st century become so specialized they become foreign to each other. Doctors consort with doctors, tradesmen consort with tradesmen, rich with rich, and poor with poor. Civilization overwhelms those small, tight-knit familial group bonds. It creates barriers between other groups because of specialized conformity. It creates subgroups that are alien to other subgroups. What side of town do you live in? Are you white collar or blue collar? Are you Catholic or Protestant? God forbid you're not Christian or whatever is prevalent where you live.

It becomes increasingly difficult for civilization to coalesce behind a common purpose until eventually it collapses from the dissipation of its own identity and purpose. Social entropy marks the crossroads where civilizations choose decadence and decay instead of progressive rebirth and renaissance.

Our modern civilization isolates humanity from the environment that sustains it. It also isolates people from each other. It promotes the creation of disassociated populations that overwhelm humanity's innate small group programming. Without identity and purpose societies disintegrate into pettiness and self-indulgence.

Dr. Eleanor Stirling, Chief Conservation Scientist, Center for Biodiversity & Conservation at the American Museum of Natural History said that civilization has matured so that there has been a general erosion of knowledge about life in an ecosystem and the necessity of maintaining that ecosystem. Civilization's ability to store food allowed people to harvest beyond what was needed or sustainable. It wasn't enough to have a full cupboard. What was stored counted as wealth, so people stored more regardless of what was sustainable. People gained wealth by building granaries and warehouses. Some foods were easier to grow and store than others. Favoring a few species over a diversity of species weakened the ecosystem. It also meant a decline in the variety of edible foods. Overproduction weakened the soil. New technologies temporarily allowed food production to grow, however, overpopulation and consumption would eventually catch up. The industrialization of food technology reduced the number of farmers, fishermen, and hunters. Dr. Stirling believed those changes in lifestyles further reduced the cultural connections with the environment.

Modern agricultural practices contribute more than 25% of anthropogenic carbon emissions. Over 40% of Earth's ice-free land is dedicated to agriculture. Fully 70% of freshwater drawn from lakes, rivers, and aquifers go to agriculture. Intense land use for growing things is a leading cause of biodiversity loss and species extinction. Massive agribusiness practices and industrial farming are far more damaging to soil ecology than methods used by traditional indigenous peoples. Dr. Stirling believes that a biocultural approach to agriculture contributes to effective conservation by focusing on the relationships between people, the land and the feedback between them.

Research by University of Washington Professor David Montgomery indicates that farmers have a lot to learn from indigenous people. His research into traditional farming methods has identified techniques that improve both soil conservation and production beyond the most common agribusiness practices in use today (Montgomery, Growing a Revolution 2017).

The transition to a sustainable civilization will depend as much on good leadership as it will upon knowledge. Leaders must set the goal and fully support the benchmarks to achieve that goal. Today, the world is threatened by climate change just as America was threatened by the USSR during the Cold War.

The cool head of responsible leadership will determine if we are as successful surviving an overheated planet. The path ahead will be determined by the quality and skill of the leaders we choose. At the same time, we must recognize that the tasks ahead are too complex and diverse for any single person. It requires working as a team.

The dependence upon fossil fuels to power the global economy is a security threat. Our leaders must clearly explain why transitioning to sustainable sources is so important to peace and a secure civilization.

It's elementary that climate patterns will change as more heat is retained by the atmosphere. Many of those changes are described in the *Fourth National Climate Assessment, Volume 1,* "All of the US will experience significant changes in the climate, precipitation patterns and extreme climate related events." At present, human enterprise is primarily responsible for global warming, but what if the negligence of civilization causes nature to fight back? What If Mother Nature joins the bonfire of the vanities?

For half a century, scientists feared that warming above 3.6°F (2.0°C) would force Mother Nature to start stoking her own furnaces. How would that happen? Biogenic GHGs come from natural sources like erosion, decomposition and forest fires. These are amplified by tipping points, and feedback loops. Human emissions work as a catalyst. Biogenic emission sources might begin to contribute more GHGs than anthropogenic sources. Once thought unlikely or far into the future, melting permafrost is one of the embers in nature's furnace that is already starting to glow.

Permafrost regions of Siberia are warming up to twelve times faster than lower latitudes. Alaska, Canada, and Russia have vast expanses of tundra where permafrost is melting at alarming rates. Permafrost and ancient limestone deposits store quantities of GHGs that are equal or greater than all the anthropogenic emissions since the industrial revolution. It is feared that these regions are already beginning to release GHGs in a growing feedback loop.

Heat extremes increase the risk of forest fires. Forest fires produce large amounts of Biogenic GHGs. As recently as 2018, scientists thought wildfires would not reach crisis proportions for a century or more. In the past decade, Europe, Australia, Canada, and the US have experienced record wildfires. Even combined, they pail compared to the forest fires in Siberia. The most aggressive forest renewal efforts will not recover full carbon capturing capacity for decades. It is sobering to think if that much change can happen in a decade, what might happen in the next half-century.

As I wrote these lines, the Bolt Creek fire in Washington state expanded in one week from an unattended campsite to a 15,000-acre raging firestorm that came within 400 yards of the small town where I live.

Biogenic feedback estimates were first modeled in 1988. The Trump administration defunded that research, eventually stopping EPA from developing specific techniques. Research in other countries continued, providing new monitoring methods and analyses. Their findings suggested that a global biogenic tipping point from multiple sources could occur in the next one or two decades. In the late summer of 2021, Sweden found biogenic emissions in their country had already exceeded anthropogenic sources. If that is an indication of a global increase in biogenic emissions, global warming may be far worse than the most pessimistic models indicate. This finding and other recent international research emphasizes the absolute necessity of increased international cooperation.

The study of global bio-geochemical systems is a relatively recent activity. The first International Geophysical Year was 1957-1958. Scientists from around the world began the first rudimentary observations of our global environment and the systems that support it. It immediately became clear that humanity was causing massive damage. By the late 1960s, activists began speaking out. Gradually protests became global, political, and socially disruptive.

In 1988, the World Meteorological Organization (WMO) and the United Nations Environment Program (UNEP) put together an Intergovernmental Panel on Climate Change (IPCC) to study and collect the most reliable climate information from international experts around the world. The IPCC soon calculated that warming beyond 2.7°F (1.5°C) would place our entire global civilization in jeopardy by triggering biogenic and environmental changes. Those changes would alter the planet's climate to such an extent that civilization itself would be threatened. As a precautionary measure in 2015, one hundred ninety-five UN members pledged in the Paris Accord to make their best effort to reduce CO^2 emissions 30% by 2020.

A reduction did not happen and global emissions continued to increase. Political and special interest influences at subsequent climate conferences simply moved the goal posts. In 2018, one hundred ninety-three nations pledged to reduce total emissions 40% by 2030. Again, nothing happened. Pledges weren't kept. Only a handful of nations made an effort to honor their pledge. Youth activists around the world began organizing and holding massive demonstrations to hold nations accountable. In the late summer of 2019, sixteen-year-old, four time Nobel Peace Prize nominee Greta Thunberg addressed the Congress of the United States. She submitted the 2018, IPCC AR5 report as her evidence. She told Congress not to pay attention to her because she was only a child. Congress should pay attention to the scientists, their own agencies and the IPCC report.

The need for immediate global action was imperative. The science was too well established to ignore. The chemistry was irrefutable, and the numbers were getting worse every day. There were two realities that were at odds with each other. There was the combination of economics and politics, and there was the bio-geochemical reality of the carbon budget and science. Conservative politicians didn't pay attention to that reality. Their fossil fuel financed political reality only delayed action and made decarbonizing the economy more urgent and vastly more expensive. Science was losing and Earth's carbon emission budget was approaching bankruptcy.

The carbon budget is the bottom line to the entire climate change crisis. Experiments under meticulously controlled conditions long ago established that 'X' amount of carbon dioxide would yield 'Y' amount of warming. If you increase 'X', you will increase 'Y'. The Earth warms more slowly in the real world because the planet has many systems that keep temperatures within the narrow limits we humans find comfortable.

For twelve thousand years, that was a good thing. Earth systems provided a resilience to human emissions until the industrial revolution. Then we started digging up two hundred- and fifty-million-year-old carbon and started smoking up the place. The real trouble began in the 1950s and the Great Acceleration of industrialization after World War II. Like adding too much heat to a pot of stew, Anthropogenic global warming threw Earth systems out of balance.

When the industrial era began, CO^2 concentration in the atmosphere was around 280 ppm. Today it has reached over 424 ppm. Human enterprise was emitting more carbon dioxide while we were simultaneously destroying the Earth systems that mitigate excessive warming. Where is the sense in that?

At the end of 2021, most estimates said we had roughly three hundred fifty billion tons (~350 gigatons) left in the carbon budget before we exceeded the warming limit of 2.7°F (1.5°C) set by the IPCC. The global annual CO^2 emission rate for 2021 was estimated to be around 41 gigatons. We can determine the approximate time left before we exceed 2.7°F (1.5°C) with simple arithmetic.

Take the remaining carbon budget of 350 gigatons and divide that number by Anthropogenic gigatons emissions per year. The answer will tell you how many years we have left before there is enough CO^2 to warm the atmosphere beyond 2.7°F (1.5°C). The answer was 8.5 years. Natural resilience systems might normally buffer that number and give us a few extra years. But civilization and a consumption-based economy has severely damaged many of those systems. Earth's natural resilience can no longer compensate for civilizations heavy boot.

It is unlikely that many nations would be capable of reducing emissions 100% in eight and a half years. If all nations could meet their pledge for a 50% reduction by 2030, that would still not attain the 2.7°F (1.5°C) target. There are three reasons for this: The first is that CO2 is not the only GHG civilization is emitting. The GHGs already in the atmosphere will continue to warm the atmosphere for hundreds of years. The last reason is that biogenic sources are now beginning to contribute significant amounts of GHGs. These three sources increase the total GHG emissions to nearly 52 gigatons per year. Bill Gates is investing billions in sustainable energy because he agrees with these numbers and the lethal threat they pose. (Gates (2021) *How to Avoid Climate Change Disaster*).

If we crunch these new numbers, we find the following: divide the carbon budget of 350 gigatons by 52 gigatons per year. We get 6.7 years. None of this arithmetic includes tipping points or feedback loops that would drastically alter the calculus and reduce the window for meaningful action even more. Global energy needs and consumption are expanding exponentially. Civilization is literally destroying Earth's resilience to human abuse through pollution, acidification, deforestation, and an economy based upon unbridled consumption and waste. Earth's resilience and adaptation portfolio grows smaller each year. Estimates of environmental and subsequent economic collapse may vary a few years one way or the other, but which figures should the world act on?

The precautionary principle suggests the world should set policy based on the worst-case scenario, not the best. For all practical purposes, the 2.7°F (1.5°C) ship has already sailed. We are rapidly approaching the juncture where the cost of environmental damage will begin to exceed the funds and resources available to maintain a secure civilization. That point is 3.6°F (2.0°C). To remain below the new target will require a net global reduction of at least 50% of all GHGs by 2030 and net zero a decade later.

When the target was 2.7°F (1.5°C) set in 1988, the global cost to the economy would have been in the hundreds of billions of dollars. The World Monetary Fund and the World Bank recently estimated the cost will be in the many trillions of dollars to reduce emissions enough to remain below 3.6°F (2.0°C). The utter stupidity of delaying longer is beyond rationality.

Chapter 9
Bugs on the Windshield

Politicians around the world seem to have missed the Economics 101 point that the environment provides 100% of the raw materials for civilization. Ecosystems recycle those materials in the process of living, dying, decay, and renewal. This is only a subset of a giant planetary system where organic and inorganic components have interacted with each other for at least four billion years. That give and take evolves on time scales ranging from one generation of a bacterium to the slow movement and erosion of tectonic plates. In between the breaking down, building up and breaking down again are infinite bio-geochemical systems that exchange matter and energy in an ever-changing dynamic equilibrium where time is irrelevant, and nothing is wasted. The evidence of the interdependence of that inorganic/organic system surrounds us.

In the warm coastal waters of western Australia, there is a slimy looking rocklike formation made by cyanobacteria (blue-green algae). Cyanobacteria were among the earliest life forms on Earth. The primordial atmosphere was hot, toxic, and too unsettled for life as we know it. Cyanobacteria was the first organism with the ability to capture energy from the sun through photosynthesis. The oxygen produced by photosynthesis was as toxic to the primordial environment as sulfuric acid would be to us. Photosynthesis and the production of oxygen was responsible for the planet's first mass extinction and the formation of thousands of chemical oxides.

Eventually cyanobacteria multiplied and spread over several billion years. Cyanobacteria was no longer alone. Green plants evolved and flourished. Photosynthesis gradually terraformed the Earth to the oxygen rich planet we have today. For about three billion years, simple organisms transformed the land, water, and air. About a billion years ago nature began vigorously experimenting with new forms of life.

A little over five hundred million years ago (Cambrian Epoch) there was another huge change. More complex organisms began to develop in the seas and on the land. In only a few million years, nearly every plant and animal group began to appear on the scene.

Soft tissue was supported by shell, cartilage, and then by bone. Bone soon formed to protect vital organs and support a spinal cord. Some plants evolved woody parts to become trees and form vast forests. Chordates (animals with a spinal cord) and vertebrates (backboned animals) adapted to life on land, sea, and in the air. Some animals ate plants, and some ate other animals. Meanwhile some clever plants invented sex.

The invention of sex was nearly as important as altering the atmosphere. It provided a more precise ability to adapt to changes in the environment. To help with the process some organisms got an assist from other organisms. A mutually beneficial symbiotic relationship (mutualism) between flowering plants and pollinating organisms flourished.

Much of the human food supply depends upon the relationship between plants and pollinators. That relationship is now in peril because the natural environment is changing so fast. Instead of major changes taking place over millions or billions of years, enormous changes are taking place in a few decades, even a few years. Instead of blue-green-algae changing the world, we are changing it.

Our influence is causing the winter to come later, and spring to come earlier. That isn't as good as it sounds. The relationship between flowering plants and pollinators is getting out of sync. The longer warm season provides an environment that is more suitable to molds, bacteria, and life forms that prey on pollinator populations. To compensate for this problem, farmers apply chemicals. Over three million tons of pesticides are applied globally each year. Many of these chemicals are known to kill entire colonies of bees and other pollinators.

Washington State produces over half of the country's annual apple harvest, as well as being a major pear and cherry exporter. Apple production alone contributes approximately two and a half billion dollars to the annual state economy. Orchardists along the Columbia River complain that their crops are in trouble because colony collapse has decimated the native bee population. There simply aren't enough bees to pollinate their orchards. Importing or buying new bee colonies has become a new annual overhead expenditure. Bees aren't the only concern. Insects play crucial roles in the global ecology by filling niches in nearly every part of the food chain.

I first became aware of the general decline in bugs while driving through eastern Washington State. Farms cover hundreds of square miles of rolling hills with wheat, soybeans, and orchards. Those amber waves of grain were a photographer's dream, except for the bugs. In the 1970s and 1980s, I would have to pull over every twenty miles or so because my windshield had become an insect killing ground. My camera's viewfinder was often filled with more buzzing bugs than scenery. I recently drove through the same farmlands without having to stop once. Where were the bugs?

The total mass of bugs exceeds that of all vertebrates combined or about seventy times that of humans. Birds, bats, and many other species, depend on them for food. There are over three hundred species of plants that feed on insects. Insects play a key ecological role consuming and being consumed.

In the 1960s, entomologists began to suspect that a decline in some insect species might indicate a trend. At the time, there were no long-term studies to establish if that was true. Researchers began collecting samples of multiple insect species and logging each year's totals. By the 1980s, the trend lines were taking on an ominous shape. Bug numbers were declining, a lot.

Recent insect population studies around the world have more clearly defined a trend suspected over half a century ago. Declining insect populations have placed the world on a path to 'ecological Armageddon', according to several studies.

Since 1989, researchers from Raboud University in the Netherlands documented insect population decreases in nature reserves across Germany. Nature reserves were chosen because humans would have the least direct influence on insect numbers. They found that there was a decline of 75% over the twenty-seven-year study. During the summer months, the decline averaged around 82%. According to global monitoring data of 442 species, there was a 45% decline in invertebrate populations over the past forty years (Dirzo, Science, 2014).

I became curious about what would happen in regions where human influence was higher. Research in this area has recently gained importance. Species-specific studies showed similar losses. In Europe as well as the US, a 30% to 40% reduction in wild and managed bee populations had been documented. A decline in some of the most recognized species of butterflies was showing up in peer-reviewed journals and the media. Many rare and protected species showed dangerous rates of decline.

A 2014 study published in the journal Science documented a steep drop globally in insect and invertebrate populations. Rudolfo Dirzo, an ecologist at Stanford University, said, "Although invertebrates are the least well-evaluated faunal group within the International Union of Conservation of Nature (IUCN) database, the available information suggests a dire situation in many parts of the world."

The gob smacking drop in insect populations has a far-reaching consequence for humans and tens of thousands of other species. For example, larval flies (maggots) play an enormous role in recycling dead and decaying organic matter. Putrefying carcasses would soon carpet the landscape without these vital scavengers.

The relationship between insects and birds are closely linked to human health. A 2010 study by Canadian biologists suggested that insect-feeding birds in North America had suffered more pronounced declines in recent years than seed-eating species. This was significant because insect-feeding birds are important in controlling disease transmitting insect populations.

Birds and bats provide much of the natural control of the female Anopheles mosquitoes (vector for malaria). Other diseases carried by mosquitoes include encephalitis, West Nile virus, dengue, yellow fever, filariasis, tularemia, dirofilariasis, and Zika. Mosquitoes are not the only insects that carry disease.

The decrease in insect populations drive a decline in amphibians and predator insects. Germany's Federal Agency for Nature Conservation identified long-legged flies, dance flies, dagger flies, and balloon flies as important predators for pest and vector insect species. Spiders are not insects, but they also play a role in controlling pest insects.

An alliance of twenty-two publicly funded environmental research institutions compiled a list of ecosystem services provided by insects. Over three-quarters of wild flowering plant species in temperate regions need pollination to fully develop their fruits and seeds.

Civilization needs pest and vector controls for food production and the protection of public health. Outdated agribusiness practices rely heavily upon the application of artificial fertilizers and pesticides. The widespread use of novel (not occurring in nature) chemical pesticides causes collateral damage far beyond the intended application. Novel chemicals kill pests and beneficial species. The profit motive drives chemical manufacturers to ignore the wider ecological role that insects, birds, bats, and amphibians play by providing the same service without charge and without toxic side effects. The agrobusiness model is designed to focus on their quarterly report and not the off-ledger system thinking necessary to maintain a sustainable global environment.

The Berlin Natural History Museum compiled a long list of factors that contributed to insect loss. When a novel chemical is introduced into an ecosystem, nature doesn't have a handy means to neutralize it. The widespread use of some fertilizers might help one plant species to thrive but prove toxic to another species or beneficial insect. Some of the other threat factors identified in the bug literature deal with climate change, habitat destruction, deforestation, ecosystem fragmentation, urbanization, agricultural conversion, and chemical pollution. Industry introduces at least fifteen hundred new novel chemicals into the environment every year. Evaluating their toxicity can take decades. Their lasting effects on the environment and species diversity remain largely unknown. Conservative politicians in the US have worked diligently to deregulate and defund research on novel chemical toxicity. I don't point this out as a partisan argument. It's just a statement of fact.

In late 2019 and 2020 swarms of locusts, described as a plague of biblical proportions, swept over the horn of Africa. The entire annual food harvest in the region was lost. This brought starvation for millions in an already drought-stricken part of the world. There may be a practical solution to cockroach and locust infestation. With proper preparation, they might go well 'with some fava beans and a little chianti'. I am not being sarcastic or cynical. Bugs are a highly nutrient-rich food source for humans in many parts of the world. I personally like my locusts (grasshoppers) fried in oil, add a little salt and they taste and crunch like corn nuts. Insects provide a common food source in many parts of the world. The problem for the uninitiated is taking that first bite.

Recent data points to climate change and chemical use as the major cause for the dramatic decline in insects, birds, bats, and amphibians. Long-term global data comparisons, however, remain scarce and in many places non-existent. The citizen scientist may offer a solution. Lay people with an interest in the outdoors can be trained to collect important information through citizen science projects. Even counting bugs on the windshield could make a contribution.

The decline in overall insect populations is just one more of the many ecological threats coming to a head in the first half of the 21st century. It is now abundantly clear that failure to respond to these challenges appropriately may result in a kind of colony collapse for our own species.

The application of system thinking is vital to understand the total influence of civilization on Earth's bio-geochemical systems. Nothing happens independently from everything else. Our use once and throw away, unregulated consumer economy is a prime example where short-term expedients have resulted in disastrous long-term consequences. If we continue to ignore this fact, that plastic cup we so carelessly threw away yesterday may eventually end up in the sushi we have for lunch tomorrow.

Chapter 10
The Paradox of Two Realities

Global warming and subsequent changes in the climate are only two of many interwoven consequences of nearsighted governance. Civilization has moved so rapidly that the average person has failed to notice the damage of unsustainable consumption. The importance of science to good governance cannot be overemphasized. Science improves civilization's knowledge and awareness of those things people can't sense. When I was training inspectors and investigators for the US EPA, I frequently described the role of the inspector as the eyes, ears and nose of the agency. If they failed to accurately report their observations, the agency could not make appropriate decisions to protect health and the environment. That same principle applies to the relationship between science and governance.

Science is simply the pursuit of knowledge. Scientific truth is based on the objective and repeatable observation of evidence gathered from many diverse sources. When we look outside, we see weather not the climate. We can't see a molecule with our naked eye, but science has proven it exists. Humans can't touch a star or measure the diameter of a galaxy, but scientifically informed technology can. We should not expect the complexities of climate change, mass extinction, and collapsing ecosystems to be visible out our kitchen window. We know they exist, based upon evidence documented by science. Good decisions cannot be made on corrupted information. Science can provide the best available knowledge to policy makers when and if they want it.

Science is not political. The current doubt in the authenticity of science must fall on corrupt governance and the politicization of scientific truth. Here rests the paradox. In recent decades, politics has taken a path toward authoritarianism and the manipulation of truth for political purposes. Science doesn't determine how knowledge is used, nor does it decide if it will be used in beneficial or harmful ways. Science has no defense when political arguments are based upon unsupportable fictions. Science cannot debate lies because there is no evidence to study or question.

In 2018, the Fourth National Climate Assessment report was released by the US Global Research Program (USGRP). It was written in coordination with the EPA, NOAA, NASA, the National Science Foundation, the DoD, the Department of Agriculture, the US Agency for International Development, the Smithsonian Institution, the Department of Transportation, the Department of State, the Department of Interior, the Department of Health and Human Services, the Department of Energy, and the Department of Commerce. The report stated emphatically that, "…changes in the environment are polluting and consuming beyond the carrying capacity of the planet. This is altering planetary systems and poses an existential threat to the US."

How much clearer can science be in the attempt to inform leaders? When governments ignore the information in thousands of authenticated scientific reports, it does so at the public's peril. The evidence points to malice and corruption within the government and powerful special interests that seek to manipulate governance. This is precisely why informed citizens, and millions of young people around the world, repeat Greta's justifiably emotional cry. "How dare you!"

A few years back I saw a cartoon that illustrated how simple the choice is. The cartoon showed a lecture hall. The speaker on stage was pointing to a screen. The screen showed a list of the benefits of transitioning to a sustainable economy: energy independence, preserve rainforests, sustainability, green jobs, livable cities, renewables, clean water and air, healthy children, etc. etc. In the audience, a skeptic leaned over to the man next to him and whispered, "What if it's a big hoax and we create a better world for nothing"?

Scapegoating science is a losing proposition in the long run. The truth always tends to rise to the surface. Does the public really doubt that extreme heat, droughts, fires, floods, hurricanes, and cyclones are a threat to global health and security? Does the public really doubt that climate change is making those events more frequent and stronger? Does the public really doubt that climate change is increasing the cost of living and doing business? Isn't it clear that the fallacious arguments against climate science are coming from corrupted politicians and special interests?

My investigation now turned to how those special interests gained so much influence? What was the other reality dominating governance?

It must be conceded that coal, oil, and gas made the world more prosperous. It is perfectly rational to wonder, if fossil fuels were so successful in the past, why change now? The counter argument is just as clear. Coal, oil, and gas are killing us. This is like the story about the man who jumped off a skyscraper. On the way down, he could be heard to shout, 'so far so good'. The man may have faith that 'so far, so good', is good enough, but I'll put my money on the established scientific probability that things will not end well. Something was persuading our government not to address climate change or protect the environment.

A group of children confronted Senator Diane Feinstein (D-CA) about the small window of time left before serious tipping-points and feedback loops placed their future in jeopardy. A tiny pre-teen girl stepped forward. She timidly looked up and asked the Senator to support the Green New Deal. Senator Feinstein ignored the child's plea. The Senator began scolding them for daring to confront her. She said they were too young to have voted for her, therefore, they had no right to ask her to act on their behalf. She would not discuss any issue or point of the children's argument. The Senator cited her thirty years of experience as proof that they should trust her. "I know what I'm doing," she said. That was difficult for the children to accept because the climate crisis had not been resolved nor had any significant steps to address it been taken. An older girl stepped forward and tried to explain that solid scientific research indicated there were only a few years left to mitigate climate change. Senator Feinstein's tone became angry. She said that instead of making demands on her to do something, the children should go back to school and solve the climate problem when they grew up. That of course would be too late to prevent the disaster they feared. This example of arrogantly talking past the public's concerns has become far too commonplace in the halls of government.

Poor governance in America and around the world is the result of the politics of power, not the priorities of the global community. It is driven by special interests and pressure from an unregulated consumer economy. Numerous polls have shown that most of the US electorate supports action on climate change, reproductive rights, universal health care, better public education, social safety net programs, day care, moderate gun control, getting money out of politics, free and easy voting, subsidized higher learning, equal rights for LGBTQ, updating America's infrastructure, and decarbonizing the economy. Very few of these public priorities are reflected in laws or policy under Republican influence. The GOP seems to be intentionally creating fear and anxiety in the general population.

We understand the need for reason, yet rational behavior has taken a back seat to more primal instincts. The divisiveness and hostility motivated by the GOP lies and distortions has destabilized the country. We need to understand how our emotions influence our behavior.

That idea was explored in the 1956 science fiction film Forbidden Planet. A father and daughter were crash survivors stranded on a mysterious planet. The planet was once inhabited by an enormously powerful race called the Krell. Their technology was so great that they were able to create and control the environment of the entire planet. By connecting their minds to an immensely powerful artificial intelligence (AI) boosting machine, they were able to create whatever they imagined. However, there was something hidden deep in their ancestral psyche that threatened this utopian existence. At the pinnacle of their power, a monster appeared that was more powerful than anything they could do to stop it. The more they used their AI technology, the more powerful the monster became until it destroyed their entire civilization. They had not realized that the monster was created from ancient instincts deep in their psyche. Its primitive destructive influence was magnified by the same technology their civilization had created.

What if the Krell had recognized the monster was a manifestation of some remnant left over from their genetic past? Could the Krell have avoided disaster? That question can just as easily be applied today. Are we advancing wellbeing and a sustainable environment or feeding the hunter-gatherer, apex predator hidden in some primitive corner of our primordial psyche? That part of the brain responsible for our basic emotions is called the amygdala.

Nigel Nicholson, Professor Emeritus at the London Business School, put it this way, "You can take the person out of the Stone Age, not the Stone Age out of the person." Is this the glitch that causes us to repeat the same mistakes from one civilization to the next? We may not have come as far from our evolutionary past as we would like to believe.

This idea came to me one warm summer day in 1969. I was in London, walking along a stone wall near Hyde Park. Dozens of artists were exhibiting their work. I came upon half a dozen wonderful pen-and-ink animal sketches. I knelt to admire an incredibly detailed drawing of a grinning Orangutang. It was almost as if he was looking back at me. His wide, cheeky grin and the sly glint in his eye gave me the impression there was something we shared but couldn't quite identify. I wondered how the artist was able to capture that Mona Lisa mystery in the expression of an Orangutang. The ape and I must have stared at each other for a full five minutes before I broke the connection and glanced up. Above the sketch was a note. It said, "This mirror, not for sale."

Indonesian people refer to Orangutangs as the red people of the jungle. Next to the Bonobo Chimpanzee, the Orangutang is one of our closest relatives with 98.8% of our DNA in common. In 1996, I spent a few hours with an Orangutang mother and daughter at a wildlife preserve near Jakarta, Indonesia. Within minutes, I began to feel a strange sense of kinship. The mother sat on my right. Her daughter was seated on my left. After a few minutes of becoming acquainted, Mom's enormous hand reached into my breast pocket and took my pen. I playfully scolded her and took it back and put it into the breast pocket opposite her. She put her face close to mine and gave me a big cheeky Orangutang grin. I soon felt her stealthily reaching behind me, to get her daughter's attention. Mom continued to distract me. When I started to look away, she began playing with my glasses. While she had my attention, her daughter deftly purloined my pen. The daughter then reached behind me and handed my pen to her mother. Now in possession of the pen, Mom gave me the same look I saw in Hyde Park nearly thirty years earlier. She looked at me, tilting her head slightly downward, eyebrows slightly raised, her mouth opened wide in silent laughter. I never saw a clearer expression of 'gotcha' in my life. I had just been grifted by an ape. The face in that 'mirror' in Hide Park was closer to mine than I had ever imagined.

Could we have created a world we are not genetically programmed to thrive in. Biologically we are not like Cro-Magnon—we *are* Cro-Magnon, trying to live in a 21st century, technological civilization. That civilization has less and less resemblance to the natural environment our DNA is adapted to. Once we faced real monsters. The most dangerous monsters we encounter today are of our own creation.

Our gene pool spread over the planet during at least three climate driven migration episodes. Neanderthals and Denisovans probably represent some of those earlier migrations, but they were all enough human to meld. The core DNA recipe was already written. It might blend and re-blend with earlier migration groups but has remained essentially *Homo sapiens* for the last three hundred thousand years.

What if a child born fifty thousand years ago, could be transported to this century? It would almost certainly have the same mental and physical tools to succeed as a child born today. In fact, it would probably be more robust and perhaps a little brighter than its modern peers. That's because modern health care has removed many natural culling pressures. The apex predator that once saw a spearhead in a piece of obsidian would now have the where-with-all to build a hydrogen bomb.

In the 21st century, that neolithic child would no longer look at the moon and make-up stories about what might be there. She would go there to find out. She wouldn't count on her fingers or with knots on a piece of string. She would use calculus and quantum physics to measure the universe. Instead of facing her enemies, she would study them from half a world away, and vaporize them with the touch of a button. This raises the question; can we raise that child so that the monsters of her world do not surface in this one?

The questions and paradoxes on the influence of nurture versus nature have filled libraries. With all the knowledge of science and history, we still choose to ignore how our innate patterns of behavior influence the world we have created. Regardless of the power of our intellect and inventive imaginations, we still fight the monsters created by our Stone Age perceptions and motivations. Can we nurture a caveman not to be a caveman?

We all inherit a genetic mental and physical envelope of potentials from our parents. Those potentials can be nurtured or starved. How those potentials are expressed depends almost entirely upon when and what experiences we have as we mature. Are we shaping the future society to be rational diplomats or nurturing a shoot first, ask questions later approach? Race, ethnicity, class, or gender have nothing to do with the human potential. How the human potential is nurtured has everything to do with it. Do the experiences of modern 'civilized' life enlighten us to be more rational and contemplative or are they crafted to manipulate and magnify innate responses already ingrained in our amygdala? What if civilization insulates us from reality and the consequences of our actions? Does it make us better or actually bring out the baser aspects of our nature?

In most respects, my father would be described as a good, brave, and decent man. He was raised by loving and nurturing parents. After returning from World War II, he struggled with the knowledge that the bombs he dropped from his B-17 killed hundreds, if not thousands. Even good people can become insensitive to mass murder and the suffering of others. War made him impatient with people who thought only bad people could be made to do bad things. Were the Germans he bombed unredeemable monsters or simply programmed by social myths, and emotional manipulations of a psychotic despot? Was there something within us that allows us to be manipulated and abandon the ties of community?

Every civilization before us, no matter how inventive, wise, or powerful eventually failed. Genetically they were all exactly like us. We would be committing a horrendous blunder of hubris to imagine superiority to any of those who came before. No civilization thus far has been able to conquer the genie in the DNA bottle. They were and we still are Cro-Magnon living in worlds of own creation. The pettiness, lies, and intrigues of today, only serve to illustrate the hurdle civilization must overcome to save itself. The cantankerous ineptness demonstrated by so many leaders reflects something lurking in the depths of the human gene pool. The beast is fed by privilege without consequences or responsibility. Excessive privilege without consequences erodes the foundation of civilizations. What is necessary to keep the beast at Bay?

Global warming isn't our fault individually, nor is the solution our personal responsibility. We are members of a global civilization, consisting of many cultures. As individuals we respond to a social hierarchy. We conform to the norms of the society we live in. We obediently follow our leaders, even when their delusions lead us astray with little regard or sense of responsibility for their actions. How are we to set policies for a harmonious and sustainable civilization on a finite planet without a deeper understanding of our own nature and the consequences of our choices? How are we so easily manipulated to obediently follow delusional leaders who direct us to perform atrocities against our best interests or personal moral code?

Chaos and fear stimulate the amygdala and release fight or flight emotions. Our innate paleolithic genome is overwhelmed by the pace and complexity of modern life. In times of crisis and chaos, humanity seeks leadership to bring order. In the chaos of 21st century America, people can't tell truth from fiction in the news or on the world wide web. Conspirators, liars, and charlatans run rampant. The AI of Twitter and social media provide the platform for the manipulation of truth and reason. The dark side of social media has allowed a small minority to disproportionately express their conspiracies and flights of paranoid fantasy. Sociopathic predators seek prey by gaslighting the public to feed the monsters of their neuroses. Truth and reason fade in the fog of befuddlement. Like children set free in a candy shop, we gorged and stuffed our pockets to overflowing. We did this because our leaders told us it was progress. The result was that nature no longer nurtures us with the gentle hug of the Holocene. Humanity is once again at the inflection point that will determine a renaissance or collapse.

A significant portion of the stress and alienation comes from civilization itself. Collectively we dammed the rivers, leveled the mountains, and cut down the forests. We homesteaded the best land and then turned it into parking lots. We sucked the life out of the soil. We polluted the ocean with plastics, feces, and fertilizers while failing to notice or admit that the environment and humanity are intertwined.

The grinning ape in the mirror is us. America has become decadent. So far, the consequences of self-indulgence have been few. We have grown to believe human exceptionalism is the natural order of things. We believe we are above nature's reproach. In truth, we've been caught with our hands elbow deep in nature's cookie jar and we resent our comeuppance. It is this narcissism that blinds us from nature's reality.

Subconsciously we know something is wrong. We stand naked with stone axes in Times Square. The speed of change confounds our personal sense of security. Our anxiety compels us to simply hang on, too frozen in fear and frustration to take bold action. We seek answers to questions we don't know how to ask. We desperately cling to the familiar and quaint, instead of adapting to the unyielding storms of change that rapidly envelops us. We feel alone and lost in a strange land. The scale of anxiety has become so great that our sense of continuity is breaking apart. We hear the alarm, but we don't want to listen anymore. The fantasies are becoming too real. We desperately want a new story, a better story, a happier story. The loss of community broke the strength of acting as a collective. Around the world people began to seek authoritarian leaders instead of their neighbors. Some leaders were raised in such sheltered bubbles of decadence that they seldom experienced the consequences for their actions.

America was stunned when COVID hit our shores. Our economic, health, and security systems were totally unprepared, despite numerous early warnings by the World Health Organization (WHO) and the Centers for Disease Control and Prevention (CDC), President Trump blamed China. In the larger picture it didn't matter if China withheld information about COVID or not. By December 2019, it was already a clear threat—it was already here.

Instead of taking immediate action, President Trump canceled national pandemic contingency plans and disbanded the very organization intended to address a breakout disease. He denied COVID was even an issue, precisely when the CDC recommended sounding the call to action. Trump's negligence gave COVID the head start it needed to spread. He said the pandemic would simply go away like the flu when the weather warmed. He condemned the science. He claimed it was motivated by partisan politics. Trump continued to call COVID a hoax and actively blocked any meaningful response.

The media had a field day broadcasting the finger pointing vitriol. "If it bleeds, it leads." The Fox News business model kept the focus on the political infighting, and not on the disease. In the next weeks over thirty million jobs were lost, and the consumer economy took a nosedive not seen since the Great Depression. The repercussions swept around the world like a tsunami.

The locks were taken off US emergency reserves only to discover that they were woefully inadequate to the crisis. Heated congressional debates had been over million and billion-dollar projects. Suddenly Congress was writing checks for trillions, just to prevent a total collapse. Our instincts and human senses failed us. Trumpian conservatism saw 'kung-flu' and not a lethal disease or a global pandemic. The new Trump GOP did not listen to the science. A political decadence was in direct opposition to the real events unfolding before it. Trump continued to remove or defund national safety nets intended to ensure public wellbeing and security. He directed more money to prop up big corporations and the tycoons of Wall Street. The powerful and rich were further separated from the harsh reality and vulnerability most Americans faced every day. Ensconced in his gold-plated bubble, Trump was intellectually, socially, and economically oblivious to America's vulnerability. He could not believe he was capable of making a wrong decision. He knew more than the doctors, generals or economists because no one had ever dared to tell him differently.

When the pandemic hit, half of the US population made less than fifty thousand dollars a year. Tens of millions lost their jobs. Without a job, those millions lost their resilience to the economic shock the pandemic presented. The Trump administration vigorously resisted widespread testing. Some of the newly unemployed could not afford to go to a doctor if they became ill. Without testing, medical experts didn't know who had COVID or where it was spreading. Experts from the WHO and CDC pleaded to increase testing. Mr. Trump's response was to withdraw funding for the WHO. He threatened technical experts with termination and attempted to cancel the Affordable Care Act. The fanciful bubble of his reality did not include a pandemic, even if it would cost more US lives than all the wars in the past one hundred years.

The party of Trump still refused to listen to the unhappy story scientists told. They refused to believe the economists that pointed to the gap between real prosperity, and the prosperity of the few. Reality at 1600 Pennsylvania Avenue was Wall Street, not massive unemployment on main street, or a disease forty times worse than the flu. What used to be the GOP, blamed it on the 'Democrat Party'. Trump told the public that his party was the happy story party of patriotism, Make America Great Again (MAGA), and unrestrained personal freedom. It was not that socialist nonsense of *e plurebus unum*. Who was the public to believe, the happy Trump party stories, or the evil democrat, socialist, communist, bolshie, tree hugging, environmental alarmists telling unhappy stories? America's amygdala was beginning to glow with the embers of emotional outrage.

This all seemed so blatant and childish to me until I played back the tapes and reviewed the news casts. It really did happen, just like that.

On the Capital steps, right-wing climate propagandist Marc Morano told people not to believe the evidence. Big Money shills cherry-picked data to promote disbelief in scientific facts. A new pandemic of political hubris, lies, and hypocrisy was spreading. The US national debt would increase 25% under the Trump administration. America wasn't told that. The Trump administration would leave the consequences of negligent hubris to the next administration and America's children.

Since 1950, and the Great Acceleration, civilization has danced on a knife edge with less and less room for error. Wealth in the US filtered up to a minority of astronomically rich. America continued consuming an unfair share of global prosperity. In 2022, the US population of three hundred thirty-two million was less than 4.5% of a global population of eight billion, yet contributed at least 15% of global pollution, 40% of the world's waste, and 25% of global energy use. America continued to build its wealth on an economic policy of unbridled production, consumption, and waste. America was so rich it became sheltered from the consequences of decadence. America believed it was the shining city on the hill, exceptional and infallible, perhaps even divinely blessed.

According to The Global Footprint Network, the Potsdam Research Institute, and half a dozen other groups that monitor humanity's ecological footprint, by 1970 humanity had already exceeded the Earth's carrying capacity. If the per-capita consumption of eight billion only matched the footprint of the average Italian, it would still exceed Earth's carrying capacity. The Internet has shown the world how the wealthy nations live, consume, and eat. Everyone wants their piece of that pie, but there is only so much pie the Earth can serve. The extravagance of exceptionalism is quite literally destroying the future. The ultimate cost in lives by a consumer economy is many orders of magnitude beyond the six million that died in the Holocaust, or the genocide of America's indigenous people, or slavery, or the deaths by all the combined tyrants of history.

Some conservative politicians call economic inequality Social Darwinism where the fittest naturally rise to the top and deserve the wealth. Ayn Rand wrote the neo-conservative playbook. She preached that selfish individualism was the key to success. The Social Darwinists and the proponents of Ayn Rand condemned egalitarian democracies. Ayn Rand and Social Darwinists failed to recognize the biological reality of humanity's innately driven social behavior and need for community. We stumbled through COVID but survived even more fractured than before. Will rugged individualism and divisiveness help us address climate change?

Our current BAU trajectory will take the US to 7.2°F (4°C) or more by century's end and keep on warming if we don't stop all GHGs immediately. If BAU is allowed to happen, our political leaders will have committed capital crimes against future generations as devastating as nuclear war. A BAU pathway is that serious, according to the 2018 US National Climate Assessment, 2021 UN IPCC report and the World Bank publication 'Turn Down the Heat: Why a 4°C Warmer World Must Be Avoided'. Our global civilization is marching down the road to Futurecide. How did normal, good and kind people come to this?

Most of us are familiar with the fable about two frogs. Which one survived? The one tossed into boiling water survived because it immediately sensed the problem and escaped. The other frog didn't notice until it was too late. The fable applies to *Futurecide* because the older generation failed to notice the changes taking place. Young people already find themselves in hot water and want action immediately. This dichotomy tends to politically alienate one generation from the other. The faster Earth's habitat deteriorates, the more that alienation grows.

The novel *You Can't Go Home Again* by Thomas Wolf was published in 1940. The theme explored the changes in American society in the nineteen twenties and thirties. He wrote about the exuberant prosperity of the twenties through the stock market crash and the agonizing recovery afterward. The economy just before the Great Depression was a great deal like America before COVID struck. When times are good, people believe they will get even better. When times are bad, people want to believe the good times are just around the corner. Older generations cling to memories of the past and dream that world will return to those 'good old days'. "You Can't Go Home Again" illustrated that time and change passes in a way that makes it impossible to '…go home again'.

It is more than the loss of clean air, clean water to drink, and rich soil to grow food that threatens to kill us. This time the entire planet is beginning to cast us aside. A collapsing environment affects every aspect of survival and any hope to rebuild. Destroying the environment means we lose everything that supports life itself, the soil, the water, the trees, fish in the sea, even cool shade and the air we breathe.

The older generation, my generation, is stuck in a rut. Changing behavior is threatening. The necessary changes are met with denial or hostility. Young people don't see endless prosperity in their future. Their portfolios are empty. They see deserts, storms, disease, and the chaos of millions and millions of humans rending each other apart like some nightmarish Hieronymus Bosch painting. That vision is not an illusion to the young. It is a clear-eyed vision of things already happening and getting exponentially worse. Young people know climate change is eating away at their future. The cost of extreme events is digging deeper into the general budget every day. Young people know that they will get the bill.

It is likely that America will spend more in this decade on damage and recovery from extreme climate events than the annual operating budgets for the Department of Education, Department of Interior, National Park Service, US EPA, Smithsonian Institution, National Science Foundation, Department of Transportation, NASA, NOAA, the Department of State, and the Department of Energy combined. No amount of juggling the numbers will cloud the truth forever.

If we act now, this instant, we can do this. We have something no other civilization in history had. There is greater access to information than all the libraries of the ancient world. We have better science and technology. More of us are educated. Our communication is global and nearly instantaneous. There is every practical reason to be optimistic. So why are we so late to get off the starting blocks? In that respect our history isn't that different from all the others. We are choosing to fail.

Most of the great civilizations passed through similar development phases. Somewhere between greed and decadence civilizations always made a wrong turn. In these early years of the 21st century, America appears to be moving from the age of wisdom into an age of greed and decadence. If we look back objectively, we can see the changes that took place and the decisions that created the world we now live in. It has never been more important to take stock and carefully chart the path ahead.

Democracy is a team sport. We err in believing any single leader is so remarkable that they alone can address the scope and complexities of the world we live in. Civilization requires teamwork. We hear debates about the priorities of money, the economy, and the necessity to maintain material growth. Seldom do we hear about public wellbeing or sustaining the natural world. People must use informed caution when picking their leaders. We must not allow lies and deregulation to blind us to an economy that filters wealth to a tiny few. Once leaders are chosen, it is up to the people to make sure they exercise their authority wisely.

It is beyond ludicrous to hear Senator Rand Paul (an eye doctor) arguing virology with Dr. Fauci (a veteran virology expert), or Senator Ted Cruz lecturing specialists from the Sierra Club, NOAA, and NASA on how to interpret their own satellite data. This is the epitome of the Dunning Kruger effect. International problems require international cooperation based upon authentic evidence.

We already have two enormously powerful technological allies. Most people don't fully recognize their potential, or the critical need to control them. AI is an incredibly powerful friend or a dire enemy. The scope and speed of AI is unlimited. AI allows us to record and store all that is known and make it instantly accessible anywhere in the world. It allows science to solve unimaginably complex problems. AI can help us manage humanity's footprint within planetary boundaries as well as help us control our more primitive impulses. It can also be used against us by stimulating those primitive impulses. We must use it cautiously because we have given AI freewill and an imagination.

Marketing research has revealed that AI already knows more about us than we know about ourselves. It has already been used to manipulate our emotions, our fears, and our appetites. We leap to action when our cell phones ring. We walk across the street when the sign says 'walk'. We easily stop thinking when AI thinks for us. That is the blessing and the curse. Will the machine serve us, or will we become slaves to the machine?

Another game changer is CRISPR (Clustered Regularly Interspaced Short Palindromic Repeats). CRISPR allows us to manipulate genes, create life, change it, or destroy it at will. In every traditional respect, these tools give humanity the power of gods, or make us subordinate slaves that walk only when told to walk or react like Pavlov's dogs when the cell phone rings.

"The first part of wisdom is knowing where to find knowledge" (~Anonymous). We are Cro-Magnons in pinstriped suits. Civilization tends to covet all that glitters, while ignoring the mundane and practical. The result is that wisdom seldom guides mankind's genius for innovation. If humans were more sapien (wise), we might be more inclined to focus our energies on social issues, the environment, and a sustainable economy. A multi-disciplined review of human health and environmental data suggests it is likely that eight in ten will die prematurely because of environmental degradation. Is civilization so distracted by the shiny objects of consumerism that it forgets what is necessary? No economy can survive if the goods and services provided by the environment are no longer available. And that's the point.

The recent pandemic was a small taste of gargantuan changes already taking place. Increased warming stirs the petri dish of virulent communicable diseases. In a way, climate change even magnifies the threat of nuclear war as nuclear powers become more desperate for resources. In August 2022, a third of Pakistan was submerged in floodwater. India and China continued to suffer once in a thousand-year floods and droughts. Their crops wither and die from record temperatures. These nations depend on melting glacier water from the Himalayas. When those waters disappear, three billion people go without food or water.

Sir Arthur Conan Doyle's character Sherlock Holms said, "Once you eliminate the impossible, whatever remains, no matter how improbable, must be the truth." Giant corporations, demigods, and politicians are not solely responsible for a collapsing environment. What remains is civilization itself. Civilization creates the specialized and sheltered lives that are no longer aware, or care about what's happening around them. Civilization tends to amplify primal instincts with little feedback to temper its hubris. On a global scale, a tendency toward self-centered nationalism and might-makes-right policies breaks down international cooperation.

After two world wars, global leaders attempted to establish a venue where diplomacy and rational negotiations could temper these tendencies. The League of Nations failed but the United Nations continues to provide a platform for the world to build a sustainable future.

In 2015, 195 UN members agreed that they could change the world for the better by bringing together their respective governments, businesses, media, institutions of higher learning, and NGOs. They identified seventeen Sustainable Development Goals (SDGs) and set a target to achieve those goals by 2030. Those SDGs provided a framework for climate action within those seventeen dimensions of sustainability. Milestones were set and one by one a global movement began to make significant progress. For the first time in twelve thousand years, all of humanity worked in collaboration to build a better, healthier, more just, and sustainable global civilization. The UN formed a global team and got to work.

The SDGs consisted of the following objectives: end poverty, eliminate hunger, ensure good health and well-being, provide quality education, achieve gender equality, provide clean water and sanitation, attain affordable clean energy, provide decent work and economic growth, encourage industry, innovation and infrastructure, reduced inequalities, establish sustainable cities and communities, ensure responsible consumption and production, attain critical climate action, protect life below water, protect life on land, attain universal peace, provide universal justice and strong institutions, and establish partnerships to achieve these goals. The US was a major leader in the movement. Every one of the seventeen goals made significant progress in the first year. Then the 2016 US Presidential election happened.

The US government quickly grew too corrupt, polarized, and intransigent to govern. The ideological cracks that had existed before, became giant chasms of partisanship. The mantra of nationalism and unregulated free-market capitalism fed the coffers of the rich and powerful, who in turn, fed the campaign coffers of corruptible and honest politicians alike. Traditional Republican philosophy morphed into Trump's personal self-aggrandizement.

Mr. Trump saw economic growth as the dominion of the 'Big Boys' of business with him as the CEO. Meanwhile, sustainably minded public policy economists like Robert Reich, Joseph E. Stiglitz, Nicholas Stern, Paul Krugman, Paul Hawken, and Jeffrey Sacks reminded us of a fundamental law of economics. *An economy cannot consume its way out of scarcity*. They found support in the research of French economist Thomas Piketty.

Piketty wrote that it was impossible to elevate general prosperity by concentrating wealth in the hands of a few mega-rich families and corporations. In his book *Capital,* Piketty documented, that it's the nature of unregulated free-market economies to filter capital to an elite minority at the expense of the majority.

Piketty's findings were sacrilege to the deregulation driven economic philosophy of Ayn Rand disciples. Hidden in the small print of the GOP platform were links to the doctrines of Ayn (rhymes with mine) Rand and her 'virtue of selfishness' conservatism expressed in her two novels *The Fountainhead,* and *Atlas Shrugged.* Big Money special interests and the Trumpian wing of the far right decided that something had to be done to silence the egalitarians, market regulators, and science. The fossil fuel corporations were not only on board with this plan, they were way ahead. Mr. Trump began disbanding science advisory boards and defunding research and oversight.

Trump placed loyalist cronies in charge of troublesome agencies like the EPA, Energy, Interior, Transportation, and Education. Trump's EPA appointee Scott Pruitt was soon removed after the EPA Inspector General reported rampant waste, fraud, and abuse of office. Trump appointed Rick Perry to be administrator to the Department of Energy (DOE). Mr. Perry was so unqualified for the task that he didn't know that nuclear energy was part of the DOE's responsibility.

Betsy DeVos was Trump's appointee to the Department of Education. She thought allowing guns in schools might keep the bears out. Ms. DeVos' brother is Eric Prince, founder of the security company Blackwater USA, now called Academi.

During America's undeclared Iraq war, Blackwater employees opened fire on a crowded square in Bagdad, killing seventeen innocent Iraqi civilians and seriously wounding twenty. Three of the Blackwater group were convicted in 2014, on fourteen manslaughter charges. In 2019, another member of the Blackwater team was convicted of murder in US court. Trump showed his disdain for the rule of law in 2020, when he pardoned them.

Trump appointed several Attorneys General to the Department of Justice (DOJ) before he found the more malleable William Barr. Trump began stacking the courts with judges selected from a short list provided by conservative supporters. Supreme Court appointments began legislating from the bench. Judiciary decisions would begin to prove more ethically situational than constitutional. The extreme right ideology was funded and backed behind-the-scenes by PAC money from the likes of the Koch brothers and Leonard A. Leo, and Barre Said.

The more I dug into these associations it gave me a sickening sense of *déjà vu* from my federal law enforcement training.

Trump ignored established norms for the office of President. His past was replete with allegations of lascivious behavior. He was under investigation for tax evasion. In June of 2023, Donald J. Trump was criminally indited on thirty-seven counts related to his illegal possession of secret documents after leaving office. His administration flagrantly and publicly ignored the laws and norms against cronyism, reprisals, lies and reprisal threats. This was standard operating procedure for Mr. Trump and many within his administration.

To me, this pattern of behavior smacked of organized crime. The more I learned, the more it seemed the Trump party (no longer recognizable as the Republican Party) was directly attacking the foundations of justice and the American democratic system of government. The evidence I've just described seemed to indicate that Mr. Trump was not acting in the public interest, but his own. He ignored the balance of power between the branches of government. From my training at the Federal Law Enforcement Training Center (FLETC) in Glynco, Georgia, it appeared to me that Trump was acting more like a mob boss Don than President of the United States.

Nepotism wasn't an issue for Mr. Trump when he appointed family members to *ad hoc* positions of authority with access to confidential information without formal title, proper security clearances, norms of protocol, or federal oversight. He received valuable considerations from foreign governments by holding meetings and having foreign dignitaries stay at his hotels and country clubs. The New York Times reported that Mr. Trump solicited personal favors from Ukraine at a time when he was under investigation. To me, Trump's policies and personal behavior seemed like the very definition of racketeering (18 USC.A. § 1961 et seq. [1970]) that I was taught at FLETC. By January 6th, 2021, Mr. Trump's public and private actions gave the appearance of deliberate sedition and incitement of insurrection to overthrow official proceedings in American democracy.

The two major American political parties began to have little relationship to their traditional platforms. Over the past fifty years, both parties moved decisively to the right. The older generation members of the Democratic Party began to appear more closely akin to an Eisenhower Republican platform than the party of Franklin D. Roosevelt. The Trumpian Strongman Party was clearly supported by Big Money, mostly from coal, oil, and gas interests. The GOP had moved to the right of an Ayn Rand novel. Was fascism rising in America?

I tracked the conservative billionaire Koch brothers. They were on the board of the John Burch Society that their father helped found. Their financial contributions lurked in the shadows, with seven figure checks to the Heritage Foundation and other neo-conservative think tanks. Republican members of the House and Senate began making personal attacks on less conservative GOP and Democratic Party members with 'alternative facts' and lies. Their ends would apparently justify any means.

In the four years of the Trump administration, the world witnessed a masterclass in the art of debating unsupportable positions by using a simple high school debating technique called Gish gallop. (Wikipedia defines Gish gallop as '…a rhetorical technique in which a person in a debate attempts to overwhelm their opponent by providing an excessive number of arguments with no regard for the accuracy or strength of those arguments'.)

The reader may recognize a more familiar definition of Gish gallop colorfully coined by American actor and comedian W. C. Fields who said, "If you can't dazzle them with brilliance, baffle them with bullshit." Overwhelm the opposition with a deluge of self-serving information and accusations so that there is little opportunity for a counter argument. Truth is irrelevant in this tactic. If caught cheating or lying, vehemently accuse your opponent of far more outrageous cheating and lying. If the opposition continues to make accusations, immediately accuse the opposition of covering up their own misdeeds.

These techniques were used again and again by climate change deniers and defenders of Mr. Trump's indiscretions. Robert's Rules of Order went out the window. Debates turned into monolog diatribes. The GOP tactic on the floor of Congress or in the media was to never let the opposition get a word in edgewise, interrupt, drown out or lie, so long as they held the floor.

Another common technique is to deflect the argument by bring up a totally different subject. This technique was repeatedly used by President Trump when he renewed accusations about Hillary Clinton's personal emails to sidestep an embarrassing topic. Those emails were usually irrelevant to the topic at hand but nonetheless stirred emotions against the opposition. Trump's arguments didn't have to make sense if it was presented as though they did.

To the Trumpian conservative agenda, the planet is simply a cornucopia of resources to be exploited, not the finite and vulnerable web science claims. The GOP would use climate change like the Hillary Clinton diversion. The Orwellian two-plus-two equals five strategy had successfully politicized science and the truth. Politics and economic policy favoring the rich was now the key obstacle to addressing climate change and sustainable development.

The crisis of climate change was turning out to be far broader and complicated than I ever imagined. If I was ever going to understand how complicated, I had to go back to school.

Over the next four years, I enrolled in forty Massive Open Online Courses (MOOCs). These were offered by some of the most highly accredited universities in the world, including MIT, Stanford, Oxford, Leeds, Exeter, University of Edinburgh, University of Chicago, University of Melbourne, University of British Columbia, Stockholm University, Columbia University, Scripts Institute of Oceanography, University of Cape Town, University of Washington, Harvard and Yale. My education expanded to economics, politics, soil ecology, agriculture, systems thinking and analysis, atmospheric physics, energy, electrical power systems, the many aspects of security, anthropology, and psychology. The more I learned the more the connections became clear. Though my knowledge was only an inch deep and a mile wide it was abundantly clear that climate change impacted and was impacted by every facet of civilization.

Climate change is irrefutable. The social, environmental, and economic components of sustainable development could not be treated independently. The paradox between politics and science could not be allowed to prevent action on an existential reality. Truth matters.

Chapter 11
Government vs Governance

The United Nations (UN) was mandated to promote and negotiate peace, equity, justice, and prosperity with all the governments of the world. The UN soon found that it wasn't the type of government but governance that determined whether a nation would be successful or not.

Good governance solves problems and ensures a safe, just, equitable, and secure society. Good governance is a prime indicator that a nation will be a successful partner on the world stage. If a government is to be successful, it will have a strong system of checks and balances. It will enforce the rules that assure equity and justice for the governed. In return, the governed must have a powerful system in place to assure that the government holds to its duty (usually defined by a constitution).

When there is corruption, citizens must be capable of bringing order back to fair play. That role was usually given to the courts that interpret the laws according to the constitution. When neither the government nor the courts rule with justice and equity, the ultimate power must fall back into the hands of the governed. This is normally expressed through the ballot box.

A small far right minority of the US population is exerting disproportionate political influence. They are experts in controlling the message. They are currently trying to control the vote. The extreme right, represented by Trump and his supporters, view free and open elections as a threat. They have increased efforts to stack the courts and place party favorites in voting oversight positions. Republican controlled states are attempting to pass laws to make access to voting more difficult for those less likely to vote for them. When the vote was not in their favor in 2010, they moved to overthrow the process and the cornerstone of American democracy and good governance.

On the other side of the aisle, the traditional Democratic Party platform of the fifties and sixties was now considered progressive extremism. The majority of the Democratic Party had moved decidedly toward the center. Traditional democratic policies became a minority wing of the Democratic Party.

With all the attention on what the system of government should be, authoritarian, fascist, oligarchy, centrist democratic, or social democracy; the storms, droughts, and fires got worse, the canoe was leaking, everyone was quarreling, and no one was bailing. This was no longer traditional politics. The GOP and a couple Democrats were aggressively blocking action to address a true planet-killer crisis. They were promoting the threat by supporting fossil fuels and blocking the possibility of a survivable future for humankind. They were destroying the environment that provides the raw materials for civilization to exist. They are literally, and willfully killing the future of life on Earth (Lord David King, head of the UK Climate Crisis Advisory Group, 2022). I began to think this must be the greatest crime ever perpetrated against humanity.

The scope and magnitude of the crisis was finally hitting me. Good governance should represent the best of social values and not the situational ethics of greed and the arrogance of ideological exceptionalism. The old guard wants to return to the good old days. Millennials want to save today. Generation-Z wants to save tomorrow. The Trump party is openly driving toward a Gish gallop fascist agenda, supported by the entire Republican Party and fossil fuel money. Why was America focusing its economy on unimaginable wealth for a few dozen people and a handful of corporations? Politicians were supporting the banks and the rich and not public wellbeing.

There is so much promise in the American people that is ready to be unleashed. Reasons for hope are everywhere but hidden by the shadows of fear and distrust. There is more accumulated knowledge, genius, and creativity today than in all of human history. If we could travel back a hundred years, our 21st century capabilities would seem so fantastic people might consider us visitors from another planet. If we went back two hundred years, we would seem like gods. We are on the cusp of discoveries that were unimaginable less than a decade ago.

Science and technology have allowed us to experience prosperity for a greater portion of humanity than ever before. We are on the verge of inhabiting other planets. Advances in scientific knowledge and new technology now provide the capability of designing, altering and creating life itself. The hope for humanity rests on one simple decision—to responsibly accept that humanity has achieved dominion over all life on Earth. We're no longer talking about saving civilization. Inaction means losing the ability to save any portion of it. If we don't act immediately, the entire epoch of man (Anthropocene) will be a tiny footnote in the pages of geologic history.

Humanity is polluting, consuming, and wasting more than the Earth can sustain. Fact! The human footprint is interfering with the fundamental systems that sustain a survivable habitat for civilization. Fact! If we don't take unified multi-national action immediately, our global civilization will soon suffer environmental consequences so severe, civilization will not have the resources or funds to recover (*IPCC 2022 report Work Group III, Mitigation of Climate Change*). Fact!

It was becoming clear to me that unregulated free-market capitalism is blind to the concept of planetary boundaries and Earth's ecological carrying capacity. No economy can last when it ignores the goods and services that feed it. Economists that understand this, refer to the goods and services provided by the environment as *natural capital*. Natural capital fuels the engines of our modern high-tech civilization. Unregulated free-market capitalism makes the mistake of considering natural capital an externality. It considers civilization and the economy apart from the natural world.

It is critically important to understand the inseparable relationship between population and the demands on natural capital resources. As the population grows, consumption and the loss of natural capital grows exponentially out of proportion. The exponential demands of a growing population were driven home to me when planning dangerous expeditions in the Himalayas.

If an expedition only had three or four 'clients', it could be comfortable with half a dozen porters, two cook staff, and a foreman (Sirdar). If there were two or three more clients, the number of porters and cook staff tripled. One year I led a group with sixteen members. It required 125 porters, half a dozen team leaders (Nikes), eight cook staff, a lead cook and assistant, two-armed security guards, and a foreman (Sirdar). Food and supplies weighed more than four tons (66 lbs./porter).

This is the same exponential reality eight billion people have on planetary resources. Every increase in the population exponentially impacts the footprint of our global civilization and the resources necessary to sustain it. The ratio between population growth and resource depletion ranges between eight to one and one hundred to one, depending upon the development level of the country. For the wealthiest nations it would be the higher figure.

The human brain is not programmed to perceive exponential change, yet it may be the most important concept affecting our future. We are not unique in this limitation. Many animals can't distinguish numbers beyond one, a couple, or many. They don't differentiate between twenty, fifty, or two hundred.

Hunters spotted by Big Horn sheep have learned that sheep can't count humans. The hunters will leave several members where the sheep first spotted them, while the shooter circles to a better firing position. The sheep won't notice that one person is missing until it's too late.

The perception of numbers in humans is only slightly different. Humans don't immediately see where an exponential change has the capability to suddenly explode our sense of reality. We can be taught the mathematical concept, but it doesn't register as we go about our daily lives. Our senses constantly lie to us. Things may seem abundant and secure today, but exponential change can alter reality completely in the next moment. We don't see it coming because we live in the immediate here and now.

Here is an example of exponential growth to help understand how it relates to our current global environment. Imagine filling the Roman Coliseum with water. It was built to hold over sixty thousand spectators, so it will take a lot of water. On day one, we'll add one drop of water. On day two, we add two drops. On day three, we add four drops, eight drops on day four, and double the drops of water every day until the coliseum is half filled with water. That will take quite a bit of time. Should we be complacent? If we applied the rule of exponential growth, when will the coliseum be completely full?

The answer is…the next day.

Exponential depletion also applies when removing water from the coliseum. On day one, we'll take one drop out, then two, four, eight, sixteen, etc., until the coliseum is once again half full. When will the coliseum be completely empty? The answer is the same…the next day.

When we look at our economy and planetary resources, we may not see much change. To the untrained eye, it looks like the world has plenty of time and resources left. The Earth is still half full. No worries, mate. Exponential math tells us exactly the opposite is true. We should panic and take action immediately to avoid disaster.

America and other consumer economies are consuming or corrupting more than is renewed at a suicidal rate. Consumer economies have not put the exponential factor into their calculations. The bank of natural capital is already overdrawn, and checks are beginning to bounce. Then comes the suffering. This is the reality of an unregulated-free-market consumer economy that ignores the exponential depletion of natural capital.

There is another dangerous factor at work. As natural resources are consumed and corrupted, the Earth loses its natural resilience to human insult. When we remove mangrove marshes, we lose tidal surge protection. This increases erosion and salt contamination of low coastal soils. When we clear-cut timber, the forest ecosystem loses a carbon sink and oxygen generator for at least one human generation. Loss of a forest removes shadow protection for streams. The streams are exposed to more sunlight. The water warms and young fish can't tolerate the warmer temperatures. Deep plowing erodes the soil. Heat and drought blow the soil away. Agricultural chemicals leach into streams that flow into rivers, that flow into the sea, creating oxygen deficient dead zones. Unregulated capitalism ignores the connections that make the planet sick. In an exponentially changing civilization, waiting until tomorrow is too late.

The four years of the Trump administration ignored the absolute necessity of protecting natural capital. His administration methodically and deliberately attacked it. His unregulated free-market view of capitalism maintained economic growth for the few by ignoring the subtractive effect of exponential consumption and degradation for everyone else. Environmental collapse has started and will be virtually locked in by 2030 if unregulated policies continue.

It is now believed that with only 3.6°F (2°C) of warming, significant portions of the tropics, Middle East, Mediterranean countries, and desert latitudes will become seasonably uninhabitable. By 2040, many regions in the desert latitudes are likely to approach temperatures as high as 135°F (57.2°C).

Let's put these temperatures and exponential change in perspective. There are no traditional grain crops (wheat, oats, barley, rice, etc.) that will grow with weeks of sustained temperatures over 110°F. It's even worse for humans. The ratio between temperature and humidity is called the *heat index*. With no wind, 50% humidity, and a sustained temperature of 109°F (42.78°C) the body cannot cool itself. The heat index is too high for human survival. After a few hours of exposure, heat stroke and death are likely. A temperature of 130°F

(54.4°C), with zero humidity is just as lethal. Exertion in either circumstance means death will occur more quickly. Humanity cannot survive without a stable and habitable environment to sustain it.

When I was with the EPA, I spent a decade monitoring corporate Environmental Management Systems (EMS). The business community understands the fundamental importance of protecting the sources of raw materials for production. If an upward curve in production exceeded the downward curve of raw material supply, production and profits suffer. If exponential change is unplanned, the shock can lead to bankruptcy.

It is an ecological axiom that populations will grow until they reach the limits of their environment. Growth must be managed within the boundaries of resources. Business management principles must follow the same exponential law of natural systems. It's simply the mathematical reality of exponential change in nature, business and civilization.

We must take a broader view of the interaction between planetary systems and human enterprise. We must look at the internet of all things. Human behavior, population, politics, economics, and environmental degradation are all drivers of global warming and subsequent changes in the climate. In turn, the climate impacts food, water, and all the natural resources vital to civilization. What may seem to be a small local change can have an enormous impact elsewhere. Unintended consequences compound over time. This is sometimes called the 'butterfly effect'.

It's important to inject a ray of hope; though the task before us is enormous, it is not insurmountable. It can still be accomplished with minimal disruption of the prosperity we already enjoy, if, and only if, we act immediately and multilaterally. Developing nations can be brought onboard at relatively low initial expenditure. Economists tell us that helping developing nations will soon result in much higher economic and security benefits for everyone. There are new methods, technologies, and near horizon breakthroughs that only need a nudge to make a sustainable transition even more attractive and profitable. Any delay exponentially increases the cost and pain of transitioning, or the inability to transition at all.

Wealthy nations have a leadership role to play. As the Earth's technological leader, how America chooses to act is especially relevant. The kind and scope of action will have wide ranging ripple effects. Therefore, it is imperative that all nations build policy together and act in unison.

The more I researched sustainable development, the more I learned how important leadership was. People follow their leaders, good and bad. If any of the major powers decided not to participate or acted in opposition, the entire climate change program could come to a halt. This was demonstrated when President Trump withdrew America from the Paris Accord. Climate action was weakened further by his assault on science and democracy. The negative impact of his actions caused other nations to hesitate. Why proceed with an enormous undertaking when the big guy on the block pulls out? Trump's disrespect for expert environmental advice was only matched by his disrespect for America's intelligence agencies.

In July of 2018, Mr. Trump and Vladimir Putin met in Helsinki, Finland. Prior to that meeting, Dan Coats, Director of National Intelligence, personally informed Trump that there was incontrovertible evidence of Russian interference in the 2016 elections. That evidence pointed directly to the top echelons of the Russian government. The assessment of Russian interference was backed by all of the US agencies and bureaus responsible for monitoring foreign intelligence activities. Those warnings were repeatedly given to Mr. Trump prior to his meeting with Mr. Putin.

The two leaders spoke privately for two hours. The only others in the room were their interpreters. Though Putin speaks English, Mr. Putin and Mr. Trump communicated through their interpreters. Mr. Trump's interpreter took extensive notes, which Mr. Trump promptly ordered destroyed. Not only was this a violation of The Presidential Records Act of 1978, (44 USC. § 2201–2209), it left the US without any record of what was discussed.

There was a 45-minute news conference at the meeting's conclusion. Members of the press asked Mr. Trump if he had discussed Russia's election interference directly with Mr. Putin. Mr. Trump admitted that Dan Coats, Director of National Intelligence and others had told him that Russian cyber-attacks threatened the sovereignty of the 2016 US elections. Trump's response was, "I don't see any reason why it would be." When a reporter pressed Trump on the topic, Mr. Trump responded by saying, Mr. Putin's denial was '…extremely strong and powerful'.

The question was put to Mr. Putin. He replied that he had been an intelligence officer (in the KGB) and understood how (fake) dossiers were sometimes created for political purposes. Putin's response implied that seventeen US agencies, and the Director of National Intelligence had created a fake story about Russian interference. It was clear that President Trump had openly accepted the word of a seventy-year Cold War enemy over that of America's top intelligence services. How would this translate with America's allies? Why should they believe America could be trusted with their intelligence or honor international agreements?

This episode became relevant to the climate crisis for the same reason. When Trump withdrew from the Paris Accord, he was acting against the strongest advice from his own government. He broke America's international Climate Accord pledge. He rejected the collective assertions of thirteen US federal agencies, including the DoD, that climate change was a threat magnifier and an existential international crisis. At that point, Trump and the Republican Party became a culpable threat to global and US security.

US intelligence agencies continued to report foreign interference and cyber-espionage up to and including the 2020 election. Many in the media, retired senior intelligence analysts, and hundreds of military experts reported that Trump's behavior was an act of complicity with an enemy nation. Mr. Trump's relationship with Mr. Putin and the activities taken by his administration did not go well with NATO members and other traditional American allies.

Madeline Albright, who served as Secretary of State under the Clinton administration, cited numerous examples in her book *Fascism A Warning*, where behavior like Trump's inexorably led to fascism and corrupt authoritarian regimes.

In her best-selling book *Strongman*, Ruth Ben-Ghiat identified how fascism grows through predictable manipulation by authoritarian leaders. She traced the methods and similarities of fascist leaders throughout history and compared them to the anti-democratic actions taken recently by authoritarian politics and Donald Trump.

Thomas Jefferson wrote that our system of democracy demands constant vigilance by the people. Benjamin Franklin was once asked, "Doctor, what do we have, a republic or a monarchy?" to which Franklin answered, "a republic, if we can keep it." When a hostile foreign leader is believed over the entire US domestic and international intelligence service, we must look sharp and ask ourselves if our republic is threatened more from the outside or from within. The erosion of democracy and the political solutions to address environmental collapse are intimately linked. A sustainable future depends upon a virtuous government not international authoritarian cronyism.

My investigations were becoming more and more compelling. If the Trump administration wasn't accepting information from official government agencies, where was it coming from? After losing the 2020 election Trump continued to distrust US intelligence that Russia had continued to interfere with elections. Instead, he blamed his loss on a conspiracy by the Democratic Party and a secret deep state cabal within the US government. He continued to support Putin's strong assertion that Russia would never interfere in US politics.

Global warming deniers argued that carbon dioxide quantities were too small to have any significant impact. Their argument that increasing the GHGs in the atmosphere by 50% was insignificant was ludicrous. It isn't the amount, but the potency of GHGs that matters. There are many instances where seemingly tiny changes in chemistry have a dramatic impact. For example, human hormones regulate growth, fertility, libido, metabolism, and hundreds of other bodily functions in parts per billion or even in parts per trillion. When a person is frightened an adrenaline release has an immediate effect on circulation, heart rate, mental acuity, and the ability to feel pain. Differences in the balance between estrogen and testosterone levels determine the expression of male and female traits regardless of their DNA assignment. A 50% increase in GHG may seem small but the effect is enormous.

Pre-industrial levels of carbon dioxide in the atmosphere hovered around 282 ppm. Prior to the industrial revolution there were few novel GHGs gases in the mix. The total load of GHGs, not including novel gases, are now above 424 ppm and growing. The addition of novel gases has become an additional concern. Some novel gases have a warming potency more than a thousand times greater than carbon dioxide. Methane was once an insignificant contributor to the mix. It now contributes nearly 40% of warming. In 2022, total GHG concentrations exceed 500 ppm. Combined, these emissions are rapidly and permanently altering the entire bio-geochemical ecology of the planet.

The numbers aren't so small when you look at the annual global contributions. There was a small reduction in the rate of emissions during the COVID pandemic. After the pandemic was brought under control, economic recovery and renewed combustion of fossil fuels accelerated emissions beyond pre-pandemic levels. At present, the emissions of all GHGs are in excess of 51 gigatons per year.

Sustainable development requires mitigating problems before they become serious. Societies must build *resilience capacity* to survive sudden stresses and maintain sufficient *adaptability* for continued change. For nearly three hundred thousand years, the human population remained within planetary boundaries.

The rise of civilization and subsequent explosion in human numbers, soon began to compromise Earth's resilience. By the middle of the twentieth century, we began to notice the change. The first international geo-physical study of planetary systems began in 1957. It revealed that human impact on the environment was everywhere. By the 1960s, federal agencies were asked to evaluate the influence of human enterprise on our social, environmental, and economic systems.

In 1988, James Hansen, NASA Director of Planetary Studies warned Congress that human influence on the atmosphere was altering the climate and posed a serious threat. In 1989, the US Global Change Research Program (USGCRP) was established. "The USGCRP is a federal program mandated by Congress to coordinate Federal research and investments in understanding the forces shaping the global environment and impacts on society. USGCRP coordinates its thirteen Federal member agencies to advance the understanding of changing Earth systems and maximize efficiencies in Federal global change research. Together, USGCRP and its member agencies provide a gateway to authoritative science, tools, and resources to help people and organizations across the country manage risks and respond to changing environmental conditions."

Under the auspices of the USGCRP the first US National Climate Assessment began in 1997. The objective was a two-part process. The first was to determine the information stakeholders needed and then to provide those stakeholders with the latest information about climate change projections for their region of the country. Numerous workshops around the country coalesced into a final document called the *First National Climate Assessment,* released in 2000. The 2000 assessment consisted of *Climate Change Impacts on the United States,* and *The Potential Consequences of Climate Variability and Change.*

It was drafted in collaboration with the thirteen agency National Assessment Synthesis Team (NAST) members. The assessment summarized and integrated the findings of regional and sector studies. It drew conclusions about the importance and consequences of climate change where it differed from natural climate variability for the United States.

Page one of The Potential Consequences of Climate Variability and Change made the following statement: "Humanity's influence on the global climate will grow in the 21st century. Increasingly, there will be significant climate-related changes that will affect each one of us. We must begin now to consider our responses, as the actions taken today will affect the quality of life for us and future generations." Since 1998, the thirteen USGCRP member agencies published more reports. Reports were routinely forwarded to Congress and the Executive Branch. Just as intelligence agencies informed Trump of Russian espionage, the USGCRP notified him of the climate crisis. Trump chose to use climate change as a straw dog and called it a hoax generated by democrats and environmentalists trying to destroy America.

The *Fourth National Climate Assessment* was published in 2018. That assessment was considerably more dire than previous reports under the following titles:

Communities: "Climate change creates new risks and exacerbates existing vulnerabilities in communities across the United States, presenting growing challenges to human health and safety, quality of life, and the rate of economic growth."

Economy: "Without substantial and sustained global mitigation and regional adaptation efforts, climate change is expected to cause growing losses to American infrastructure and property and impede the rate of economic growth over this century."

Interconnected Impacts: "Climate change affects the natural, built, and social systems we rely on individually and through their connections to one another. These interconnected systems are increasingly vulnerable to cascading impacts that are often difficult to predict, threatening essential services within and beyond the nation's borders."

Mr. Trump's reaction was to cut funding for those organizations or change their leadership to be more tractable to his desires. Despite attempts to distort the facts with the pettifoggery of Trump's political double-speak, the integrity of US government agencies remained bruised, but unbroken. Despite ethically and intellectually bereft presidential appointees, the professional civil servant staff in those federal agencies continued to honor their oath to the US

Constitution. The following is the oath I took when I joined EPA. It is virtually the same for all US Civil Servants including the President.

"I, (name), do solemnly swear (or affirm) that I will support and defend the Constitution of the United States against all enemies, foreign and domestic; that I will bear true faith and allegiance to the same; that I take this obligation freely, without any mental reservation or purpose of evasion; and that I will well and faithfully discharge the duties of the office on which I am about to enter. So, help me God."

There is nothing in that oath to suggest an allegiance to any person or political ideology above that to the US Constitution. In my personal experience, most civil servants take their oath and their professional obligation seriously and with great pride. It is still possible to search the web for EPA, NOAA, NASA, the DoD, or any of the thirteen federal agency members of the USGCRP and find tens of thousands of authoritative pages on climate change research and analysis.

The Trump administration frequently accused entire federal agencies and Civil Servant staff of conspiring against the US Constitution. He claimed that there was a secret cabal to undermine his administration and destroy America. Despite repeated official reports by the USGCRP and the thirteen federal agencies backing them, Trump claimed that climate change was a hoax, and that transitioning to sustainable energy was a plot to destroy the economy. His unsupported accusations were a nearly daily occurrence.

There are 2.1 million civilian federal employees. In my experience, it is unlikely that a cabal, hoax or conspiracy could be maintained without a massive leak of some sort. Trump's allegations were without foundation, just as his denial that Russia had tampered with the 2016 or 2022 elections was unsupportable.

According to the Daily Beast, "At least 18 people connected to President Trump have been locked up, indicted, or arrested since the real-estate mogul announced his candidacy in 2015." As of August 2022, there were at least seventeen civil or criminal investigations of Trump or Trump activities.

Is Donald Trump a "Strongman" leading a fascist movement like those described by Madaline Albright and Ruth Ben-Ghiat? The electorate of America will need to decide if that is true before their voices are silenced.

It is said that luck is when opportunity meets preparedness. How can we prepare for the future if we refuse to follow overwhelming evidence on the most pressing questions of our day? The study of the bio-geochemical components of planetary systems is challenging. Scientists who study bio-geochemical systems can't predict the precise path of the Colorado River a million years from now, any better than economists can precisely predict the price of a bushel of wheat two years from now. What science can see are trends, patterns, and probabilities. The reality of climate change demands that leaders weigh the odds and make the tough decisions. Those decisions may require unpopular choices. Sometimes those decisions must be made on narrow margins of probability and significant uncertainty. There is never perfect information, but in a crisis, lingering indecision is suicide. People can forgive a questionable decision quickly corrected, but they never forgive indecision that lingers into chaos.

Governments tend to be reactive rather than proactive. That does not serve society well in multiple, rapidly evolving crises. With adequate data, science may be able to estimate the probability of what can happen. When science reports to the government, it is up to the policy makers to determine what to do about it. There is a point when waiting too long increases the odds of a negative outcome. The empirical evidence of impending climate catastrophe is conclusive. The future cost of inaction far outweighs the cost of taking immediate action. The decision to address climate change now will save hundreds of millions of lives in this generation and billions of lives in the next.

America has a history of producing generals capable of making tough decisions at the most trying times. General Dwight David 'Ike' Eisenhower was appointed Supreme Commander of Allied Expeditionary Forces to command Operation Overlord (The D-Day invasion of Europe). Weather reports prior to the invasion would never be perfect. Ike's decision would commit hundreds of thousands of troops to the largest invasion force in human history. That decision would have to be based on the best available information. The conditions on June 6th, 1944, were not perfect, but indecision and delay would have compromised success even more.

Global warming presents a similar problem. Certainty of a result can only be 100% accurate after it happens. Decisions must be made, and action taken on the best available information and the most probable result. Policy makers must be willing to plan and act on the weight of probability. The good news is that we have far more accurate information than Ike had on D-Day. Action on climate change is not a decision where the odds are fifty-one to forty-nine.

Every credible scientific organization on Earth tells us that the odds of disaster by not addressing climate change are more like one hundred to zip. So, what options do we have?

If we take immediate action, saving most of civilization is achievable. Science and technology have given us tools undreamed of only a decade ago. The entire accumulation of human knowledge is available at our fingertips. Today we double the sum of human knowledge every thirty-six months. In a decade, it may only take a few weeks. Communication is global and instantaneous. The human potential has never been as clear or so achievable. If we look at the positive things, the future could hold, it might help us achieve it.

Abraham Lincoln is credited with saying, "The best way to predict the future is to create it." We have the science, the engineering, and the technological capacity. The real obstacle is what we believe is possible. Doing nothing, achieves nothing. My father put it this way, "Whether or not you think you can, you're right." If people don't believe it, it won't happen.

In 1987, the World Commission on Environment and Development defined sustainable development as '…meeting the needs of the present without compromising the ability of future generations to meet their own needs'. We know sustainable development rests upon three interdependent social, environmental, and economic pillars. The mechanism to build and maintain the balance between those three pillars is political, but the driver behind politics is human behavior. It isn't the difficulty of the task, but the clarity of the vision that matters. Herman Wouk wrote in *The Caine Mutiny* (1951) "When in danger or in doubt, run in circles, scream and shout." That was insanity then, just as it is today. How do we get out of the rut of constant chaos and fear? It's a little like being a dog in a cage full of squirrels, or perhaps a squirrel in a cage full of dogs. It's hard to arrive if you don't know where you're going.

Sustainable development requires changes in our cultural perceptions even more than politics. No single pillar of sustainable development can be given priority because none can exist without the other. If we are to successfully address the complex, intertwined crises the world faces today, we must meld humanity into a single force and face these issues as one people, sharing a single global habitat. Anything less and we are promoting the potential to lose half of humanity and 95% of life on Earth.

We know that some perceptions have become so firmly entrenched in our cultural psyche that they will be very difficult to change, but they must change. One of these is a system of economic growth that ignores the environmental pillar of sustainability. Tim Lenton, Professor of Climate Change and Earth System Science at the University of Exeter, narrows the problem to a struggle between two perceptions of economic reality. The current economic policy is based on the false premise that the environment is a subset of the economy. Reality is exactly the opposite. The economy is a subset of the environment. The environment would get along just fine without the human economy, but the human economy cannot exist without a healthy environment. The reason is clear.

Every new person is a consumer, but population and consumption do not have a one-to-one relationship. The Clinton administration put it this way: "The sum of all human activity, and thus the sum of all environmental, economic, and social impacts from human activity, is captured by considering population together with consumption." The rate of consumption increases from twelve to one hundred times the rate of population growth. If all other factors remain constant and the population continues to grow, sustainable development is mathematically impossible.

Despite all the pessimism, humanity has made unimaginable positive and rapid changes before. Think how the printing press changed public literacy, how the steam engine triggered the industrial revolution, AC current lit up the world and the assembly line improved manufacturing. Norman Borlaug's agricultural revolution fed millions and the transistor irrevocably changed technology. In this century, think how the Internet, computers and the mobile (cell) phone all changed the world in less than a decade. The future of AI is limitless. Each one of these altered the entire course of history and civilization.

If we focus it can be done—just imagine.

Chapter 12
Deregulation Gives Corruption License

People ask me what the economy and politics have to do with climate change. I must admit that surprises me. Here it is in a nutshell.

The economy of our global civilization consists of the exchange of goods and services resulting from production and consumption. The entire supply side of this activity comes from the environment. Production waste and over consumption have corrupted natural processes that provide the goods and services for the economy. This alters the larger planetary chemical and physical systems that must be maintained in balance to sustain the environment and the human habitat. Today the global economy exceeds the planet's ability to sustainably provide wellbeing, and security. The rapidly cascading change in our environment is an existential threat to civilization and human survival. The failure of politics to allow the regulation of the economy within planetary boundaries is why the Earth is warming.

The idea of free market capitalism was described in 1776, by Adam Smith in the *Wealth of Nations*. His concept of free markets soon began to mutate into unregulated free market capitalism, largely due to powerful business influence in government policy. Smith had warned against this pitfall. He believed it would lead to unfair competition, a decline in innovation, and monopolies. 'Unregulated' was a Pandora's box that should not have been opened.

When I taught environmental law and enforcement at the National Enforcement Training Institute (NETI) in Denver, Colorado there were three principles we drilled into each new team of inspectors and investigators. Without oversight we have no eyes, without regulations we have no standards and without enforcement we have no control. When America began deregulating economic policy, a door opened that would be difficult to close. Without oversight, controls, and enforcement, the economy would gravitate toward monopolies, stifle competition, and gain control of government policy; precisely what Adam Smith warned about. A good indicator of big business influence on government policy was the increase in the number of lobbyists infesting and investing in the halls of government.

The term 'lobbyist' first appeared in 1830. One of the first examples of corporate influence through lobbying occurred in 1850 when arms manufacturers 'gifted' custom made guns to government officials. By the turn of the century, these gifts became more generous. The public became concerned that elected officials might grow to like their relationships with lobbyists more than their constituents.

The Founding Fathers were concerned about foreign influence and addressed it in Article 1, Section 9 of the US Constitution that banned Emoluments (gifts or titles) from foreign and domestic sources. 'Gifts' were anything of value, not necessarily money. It wasn't until the 1950s that the Congress passed the Lobbying Registration Act (LRA). But special interest influence had already gained a substantial foothold. It wasn't long before the LRA was modified to allow more lobbyists than ever. It became clear that special interests, like monopolistic corporations and Big Oil money, would seek to reduce oversight and enforcement that prevented influence peddling.

Today there are more than twelve thousand registered lobbyists in Washington DC. Their influence is part of the standard operation procedure of federal and state government. No politician would write a bill without first consulting those special interest lobbyists, political campaign funding might depend upon it. The electorate is aware of this with growing cynicism and distrust.

Polls indicate that a significant portion of the population believes the government no longer has their best interests at heart. Other polls indicated that the electorate no longer believes that politicians are generally honest, virtuous, or truthful. Many believe the government favors the wellbeing of corporations and special interests.

The Reagan administration promoted corporate tax reductions and deregulation under the presumption that corporate prosperity would automatically trickle down to the masses. However, recession after recession should have made it clear that an unregulated, free-market capitalist economy is a threat to economic equity and justice. Money floats to the top and stays there. When recession hit, the working classes bailed out Wall Street and the banks. Social welfare morphed into corporate welfare.

The environment no longer had a say in politics or the economy. By the 1990s, the economy began to ignore social needs. The economy needed a growing population to consume more to keep the cash flowing through the banks to the rich.

The Citizens United v. Federal Election Commission case proved to be a keystone threat to representative democracy and human beings. That decision opened corporate 'person' wallets and granted political representation to the highest bidder. Those few that prospered under a deregulated free market economy kept the money. They could now buy more favors from politicians by contributing to campaign chests. Only a generation ago, the idea that a corporation or investment bank was 'too big to fail' would have been considered outrageous. Every Christmas we see it dramatized in the film *It's a wonderful life*.

Republican President Theodor Roosevelt campaigned against big business monopolies. In those bygone days, free market meant success or failure depended entirely upon good business management and open competition. An unmonitored and unregulated economy cannot be sustained when natural resources are threatened daily by climate change, fires, floods, storms, and droughts. The rich never see the threats from inside their gated communities. In an unregulated economy, the rich are granted special privileges in banking, taxes, and justice. Some politicians are bold enough to ask why laws and regulations are necessary for corporations. At the same time, they support more police and bigger jails for the struggling population. That's not giving equal representation and free speech to corporations—that's giving them the keys to the treasury.

When the Earth was less crowded, freedom meant something entirely different. To help understand the need for rules of behavior, imagine you are the only person in an elevator car. You are free to pass gas, curse your boss aloud, do a jig, or scratch your privates without a care, but fill the car with people and it's keep your hands to yourself and mind-your-manners. Without rules there is no equity. A rich man (Donald J. Trump) may even grow to think he can grab a woman's 'pussy' without a care. Earth is now overcrowded with eight billion people. There must now be international standards of behavior, oversight, and enforcement or the powerful will grab 'justice' without a care.

Adam Smith said, "Consumption is the sole end and purpose of all production." Recent US history reveals a growing struggle between two fundamental social philosophies. One favors privilege for an elite class, and the other favors general wellbeing. The rich aren't concerned about the working classes—it's all about the sovereign power of wealth. Unregulated free-market capitalism favors greed, unbridled consumption, and the unlimited accumulation of money. Power goes to those who control production.

Unregulated free-market capitalism claims there is no need for oversight or regulations. They claim that free-market capitalism will be self-regulating. Deregulation's core principle held that consumer demand would determine production. That was a false premise from the start. As social animals, people are easily motivated to 'keep up with the Joneses'. Marketeering firms prey on human behavior to motivate us to consume well beyond what we need or in many cases, afford. There was the assumption that markets can grow indefinitely. Resources should always be available and continued population growth will continue to consume and ensure economic growth. That was a big glitch in the system. Core resources for energy, building, and manufacturing are not unlimited in America or anywhere else on spaceship Earth.

In an unregulated free market, the gap between the rich and labor classes will always widen (Thomas Piketty). Fewer resources will mean higher prices. The purchasing power of the consumer will be stretched thinner and thinner. To temporarily solve this problem, some economists and bankers got together and decided that credit was the answer to keep GDP humming along. If the consumer couldn't pay cash, they could buy on credit. The rich will collect the interest. The banks would make their money on interest and business would continue selling with so much down and so much a month. That worked until credit limits were reached and the economy stalled (recession) again. Unregulated free-market capitalism establishes a cyclical pattern of boom and bust. Throw in a little fraud and risky speculation and you have events like a tech or housing bubble.

The wealthy will be the last to feel these shortages. They won't notice if bread is two dollars a loaf or twenty. They won't understand why the average John or Jane Doe American has a problem when an appendectomy costs $28,000.00 in the US but only $3,000.00 in Germany. Those inequities are built into unregulated free-market capitalism. The rich don't see the inequity of antiquated infrastructure or degraded water supplies so that the only safe drinking water is bottled water. The rich own the bottled water companies. This all circles back to climate change and fossil fuels.

A stable economy demands a stable energy supply. Fossil fuels can no longer provide that stability. Civilizations demand for energy increases exponentially. Constant energy growth from a finite source can neither be stable nor sustained. Demand will always eventually exceed supply. That vulnerability makes the entire global economy vulnerable. The economy will rise and fall with the cost of fossil fuels.

Another fallacy of an unregulated free-market economy is that the consumer will only buy what they need and cannot be persuaded to purchase beyond those basic needs. That's the argument that demand will always determine production. The reality of human nature is quite different. Marketing experts will tell you that it is possible to sway people to purchase far beyond their means or needs. The motivational power of marketing is so sophisticated that it is possible to sell refrigerators to Eskimos and custom Air Jordan shoes to people who can't afford food.

Over time, deregulation generates a hereditary aristocracy of wealthy families who gain power and influence over both the economy and the government. Their interests become the economy's interests. The government and courts will evolve to support them and not the majority. The laws of man will have taken priority over the laws of nature. "It's the economy stupid," is a declaration of their ignorance.

Labor is an overhead expense to production, therefore, in the business model, wages must be held to a minimum. Stagnating salaries trap more of the population in debt. Eventually debt becomes such a burden that consumption slows or collapses. This pattern was thoroughly documented by economist Thomas Piketty in his 2014 best-selling tome *Capital—In the Twenty-First Century*. Much of the trend he described was attributed to the influence of the extremely rich and multi-national mega-corporations. Piketty was not alone in his assessment as we have mentioned earlier.

Corporations are generally defined as non-profit or for-profit. In their charter, for-profit corporations clearly cite profit is their *raison d'etre*. Human wellbeing is only relevant to the for-profit corporation if it impacts their bottom line. The rich tend to philosophically and politically side with that corporate culture because so much of their wealth is tied to the stock market and investments.

Extremely wealthy families can live extravagantly on their investments and interest alone. Their money makes money. There will always be larger yachts to buy, spaceship joy rides, or $1.5 million dollar collector watches to show off. Like corporations, the fortunes of the hereditary mega-rich are managed to make more money.

The cultures of vast mega-corporation monopolies are at odds with small businesses and salaried workers who live more closely to community priorities. Small communities are decimated when a Walmart or Costco move into the neighborhood. Wholesalers give huge discounts to these enormous high-volume retailers. The cultural differences between the corporation and the rest of humanity are so great that it can only be described as sociopathic. The corporation's single goal is to maximize their profits by getting consumers to contribute the maximum amount of their income to the corporation. Conversely, the goal of the consumer is to ensure their personal security, wellbeing, and a prosperous future for their children. Those two goals are antithetical. The present conservative economic philosophy is based almost entirely upon GDP and stock market growth. It ignores the disproportionate redistribution of wealth to the rich, and the rapidly declining natural capital of the environment. Recent shifting of the political center toward the right has favored an economy that concentrates power to an elite aristocracy. This is in direct opposition to traditional Republican conservatism that abhorred too-big-to-fail banks, monopolies and championed benefits for the working classes.

Unregulated free-market capitalism tends to favor financial speculation with higher levels of risk. Those few that control most of the money make most of the money. They seek to insulate personal fortunes outside regulatory oversight. When a recession hits, it's the public that foots the bill for a too-big-to-fail, publicly funded welfare program for the wealthy. This disproportionate distribution of wealth leaves the wage earner living paycheck-to-paycheck, indentured to the wealthy through generations of income stagnation, inflation, and debt (Robert Reich, Joseph E. Stiglitz, Thomas Piketty).

There is another consequence of the widening gap between the haves and have-nots. America's national resilience was not strong enough to withstand the economic shock of a single pandemic crisis. The have-nots had a far lower resilience to sudden economic surprises. When tens of millions of jobs were lost during the first months of the COVID pandemic, nearly half of the US population had less than four hundred dollars in the bank. Nearly half of the newly unemployed were immediately broke and growing hungry. Their plight began to weigh on the nation. Meanwhile the rich 1% wondered why there were so few waiters to deliver their steaks and martinis at the country club.

Deregulation does not share economic prosperity equally. A well-functioning government monitors and regulates business and society so that they can both thrive. Deregulation and a GDP driven by consumption only works if the resources last and the people don't revolt.

Much of American economic prosperity is illusion. Incomes are higher but what that income can purchase without increasing personal debt is diminishing. There are many alternative economic theories, but lack of control can bring any of them down. Deregulation of free market capitalism has been marketed so effectively by special interests that the public perceives anything else as dangerous or unpatriotic. There is little question that overall national prosperity has increased the GDP, but the cost of living a middle-class lifestyle has been so grossly inflated that basic security now teeters on the edge of calamity.

"It was the best of times, it was the worst of times…" quotation by Dickens could just as easily apply to the gilded, city-on-the-hill image of America today. Despite the fragility of everyday existence, the public perception of America is one of unshakable wealth, international dominance and security. The perception of America around the world is quite different to anyone who travels extensively.

By 1970, America was clearly moving from a creditor nation to becoming a debtor nation. Personal debt increased while tangible equity had decreased. Fifty years later we have the illusion of prosperity but only by increasing personal debt with much lower personal ownership. This can be illustrated by the purchase of a new automobile.

In 1970, the average automobile cost around $3,600.00. The average annual income was around $9,800.00. According to Consumer Reports, by May, 2022 the average cost of a new automobile exceeded the average US per-capita annual income of $46,000.00. Prior to 1970, it would have been difficult to get more than a three-year loan to purchase a new car. The buyer would own the car outright at the end of three years with many years of useful service left in the vehicle. The remaining life of the car would count as tangible personal equity. Today it is possible to get a five, six, or even a seven-year loan on an automobile. The pay-off cost of an automobile with a six-year loan will increase significantly because of interest over the length of the loan. The final equity will be lower because the car is six years old at pay off and depreciation has reduced its market value significantly. The buyer might choose to lease a car to keep payments down. In that case, the car may be in the driveway, but the bank or leasing agency maintains the equity value of the vehicle. Personal equity on a leased car is therefore zero.

Home equity is another example where the average citizen has moved backward over the past fifty years. In 1970, it was common for a US household to have one working adult. The median cost of a new home was $11,900. If we use the same average 1970 income of $9,800, this gave a home price to income ratio of approximately 1.2 to 1. In 2021, the median cost of a US home was $347,000. The median household income was $77,000 for a home price to household income ratio of 6.6 to 1. Even with all adults working full time, a new home may already be beyond the means of many families. Personal debt is so large today that the loss of one or two paychecks can signal personal economic disaster. The picture is far worse for minority populations.

In 1984, I purchased a house in Seattle, built in 1926, for $149,000. The only improvements we made were a new roof, deck repair, hot water heater and landscaping. Today the tax appraised value is over $800,000. I could not afford to buy that same home today as a senior federal investigator with thirty years of tenure.

The income of the average American has not kept up with inflation. It should be clear that the better measure of wellbeing isn't the amount of income that matters. It is what that income will buy in proportion to inflation. Twenty-first century Americans work longer hours than their parents. Today's wages have increased but what their income will purchase is several times less. Unregulated free-market capitalism rewards the average citizen with less contentment, personal security, or prospects for the future.

Despite the policies of recent administrations, there are other economic theories or a variation of capitalism that are better suited to the 21st century. The most stable systems balance the social, environmental, and economic pillars of sustainability equally. These more progressive systems consider money only as a means of exchanging goods and services to improve security and wellbeing, not an end-in-itself. Eventually all wealth boils down to sustainable natural resources. If there is no soil and water, there is no grain. If there is no grain, there is no bread. No amount of money will make food appear out of thin air.

Not surprisingly, those nations with more progressive economies have also proven the most resilient to economic shocks, downturns, and unsettling events. We seldom hear a US politician mention contentment, personal wellbeing, or happiness when talking about the economy. Many politicians on both sides of the aisle argue that progressive policies are abhorrent to American democratic principles, but is that true? What does progressive really mean?

Here's a Wikipedia definition: "Progressivism is a political philosophy in support of social reform. It is based on the idea of progress in which advancements in science, technology, economic development, harmony with the environment, and social organization are vital to the improvement of the human condition."

On a personal level, I measure my family's wellbeing in three ways: security, prospects for the future, and happiness. Do I have security in my ability to provide food and shelter? Will my income secure my health and the health of my family? Is there enough left over for retirement? Can I educate my children to the best of their abilities? Is my children's future at least as good or better than mine was?

My family's happiness index considers other things. Do we have adequate social, recreational, and creativity opportunities? Is there time for spiritual reflection? These factors define typical middleclass aspirations. After the turmoil of the Second World War, my parents and the government ensured those things for me. What kind of world am I leaving behind? Are those aspirations even achievable today? Is the environment intact and functioning to supply my children and their children with a healthy, sustainable future? If those aspirations are unattainable, it doesn't matter how much money I make, my family's wellbeing is moving in the wrong direction. Has the US lived up to a commitment to sustainable development that meets '…the needs of the present without compromising the ability of future generations to meet their own needs'?

Your answer is probably the same as mine. The 'same old, same old' economic policy of unregulated free-market capitalism has failed miserably in achieving a truly sustainable, equitable, and just civilization within planetary boundaries. An unregulated economy removes oversight, standards and control. An unregulated free-market economy is fundamentally sociopathic and bankrupting the average American's future and the future of life on Earth.

There is another component of civilization that drives insecurity but is seldom considered by policy makers. The elephant in the room is overpopulation.

Chapter 13
Overpopulation

Overpopulation means consumption by a population exceeds the sustainable supply of resources for survival. This system operates in both directions. While the environment limits population, the population limits the environment. When a deer population eats all the forage, the reduction in forage will reduce the deer population (famine). In a healthy ecosystem, hundreds of plant and animal species may be involved in keeping the ecosystem in a dynamic equilibrium.

When the world population was smaller, humanity had a much smaller collective footprint. Planetary systems and the local ecology were less vulnerable to human insult. With eight billion people, the natural balance in the environment can no longer provide the basic resources to run a sustainable civilization. Uncontrolled population growth will negate any progress we make in reducing humanity's ecological footprint. The more the human population increases, the more planetary systems will be thrown out of balance and the more abrupt the collapse will be.

Much of the argument against women's reproductive rights, abortion and family planning rest upon arcane religious doctrine established when the global population was a few hundred million. For a religion to grow, it had to produce more followers. The modern moral issue is too many people. We must decide between outdated reproductive dogma or the collapse of civilization from privation.

Stephanie Feldstein, Population and Sustainability Director at the Center for Biological Diversity said, "Reproductive freedom is not only a basic human right. It's also a critical part of addressing human population growth and fighting the effects of climate change." We have no other choice than to alter our reproductive behavior. The extreme right agenda will make that impossible. Their resistance is not based on a moral position but the manipulation of outdated religious doctrines for political purposes. They hypocritically feign moral insult while actively fueling extinction.

We can illustrate the problem by using Buckminster Fuller's 'Spaceship Earth' analogy. When our crew launched, it had systems and supplies to sustain a voyage through space that could last indefinitely. What would happen to those systems and supplies if we doubled, tripled or quadrupled the population? Reality is much worse. Humanity has increased the population twenty-seven times more than it was two thousand years ago. At the beginning of the Holocene (12,000 years ago), the population was between one and ten million. We'll split the difference and assume a population of five million. That means we have increased the global population one thousand six hundred times what it was when humans established the first permanent settlements.

A prospective mother must be prepared to give birth and raise a child with every opportunity to prosper and develop to its fullest potential. It is immoral to be compelled to give birth and impoverish both the mother and child as outcasts of society. When overpopulation is added to the issue, an unwanted pregnancy, and the delivery of a child into a world of privation borders on insanity.

The urgency of the environmental crisis has similarities to taking a transoceanic flight. There is a point-of-no-return where the opportunity to turn back is lost. I want to make that point clear. The experts and the data clearly indicate that the point of no return for climate change is 2030. Even that is assuming there are no tipping points or unforeseen feedback loops that shut the window of opportunity sooner.

We hear the word *catastrophic* frequently when we watch the news. What does that mean in reference to climate change and environmental disruption? It means there is a point where the cost of rebuilding, maintaining resilience, and adapting to climate extremes will exceed the ability to maintain the fundamental economy and infrastructure of civilization. It means food, water and the materials for shelter will no longer be available to support the massive settlements of civilization. If humanity is going to prevent a catastrophic environmental collapse of our one and only habitat, there must be immediate, pan-national action on the three big forces driving it.

We must curb population growth. We must decarbonize the economy of all GHG emission sources. And we must transition the global economy from unregulated consumption, with few controls, to one of sustainable development within the bio-geochemical capacity of the planet. If we fail to curb population growth, the other two factors become unachievable because gains will be negated as more mouths step to the table. The choice and responsibility are entirely ours and ours alone.

The crises may be dire, but the solutions are not. Instead of an economy based upon building more we can design the economy to build better, more circular, and sustainable. We don't have to collapse the economy to accomplish this. One industry's waste might be another industry's raw material. Reuse, repair, and recycle need to become larger components of the economy. Goods can be designed so that they can be repaired, not used once and thrown away. Make things that last and not engineered for predictable failure. New businesses will pop up that do the repair, sell for reuse or recycle for use as something else.

In the 21st century, we will no longer see traditional lifetime career employment. Most people will experience several different jobs in their working lives. Decarbonizing the economy is not a job killer. It is a job changer, and an enormous job creator (Paul Hawken). Initially jobs will be taken over by robotics and AI. That is unavoidable as technology develops and replaces traditional manpower. Eventually AI will be able to replace white collar as well as blue collar jobs. Fossil fuel careers will quickly fade away, but new, safer, and higher paying jobs with new technologies will take their place. Educational support and training assistance will be key components in this transition. It will also ensure that AI remains under human control.

Change is a threat to old school thinking. When the EPA began enforcing the Clean Water Act, it was a burden to those who had not controlled their discharge. On the other side of the coin, it also created jobs that engineered and designed new practices and new waste management equipment. There were new jobs in research and engineering. Soon waste management became a huge new business. The numbers were clear—in the end, jobs and the nation benefited.

Eight billion people must be fed. After energy generation, agriculture and land use is the greatest contributor of global GHGs and pollution from chemical runoff. There is enormous potential for improvement. Decarbonizing technologies will require the reinvention of agriculture and farm labor. The soil and water crisis has become more obvious. Current corporate agribusiness practices destroy natural soil fertility and water renewal processes. Research is showing ways that will improve production, reduce a major source of GHGs, and provide more and more nutritious food (David Montgomery and Anne Bikle, *What Your Food Ate*).

Every regulation places some burden on some sector of society but if properly designed can open opportunities in others. Without monitoring, regulations, and enforcement the feral profit hogs of the economy would run rampant over planetary boundaries. Everything humans do is linked to the environment. What happens upstream has downstream consequences. A single feedlot operator might be careless with runoff. The acts of a few individuals can have repercussions affecting thousands. Pollution from fertilizer and herbicide runoff can destroy a river. A plant manager might cut corners and discharge into a storm drain. A developer might clear cut a forest and expose a stream to sunlight and sediment. Waste management is a civic and environmental responsibility in every field of endeavor.

European colonization nearly drove the American wilderness to extinction. The nineteenth century westward expansion forever altered one of the Earth's great natural ecosystems. Great hardwood forests once blanketed the east. Western conifer forests stretched from Mexico to Alaska. The Great Plains consisted of some of the deepest and richest soil in the world. Native grazing animals numbered in the tens of millions.

Bison, and other herbivores ate the grass and pooped. Their ecological niche was that of plant consumer and source of fertilizer. Microorganisms refined the manure into nutrients. Burrowing rodents and other organisms cultivated those nutrients into the soil. The nutrients helped grow the tall-grass prairie and complete the cycle of renewal for the bison. Energy and nutrients were recycled in a dynamically balanced ecosystem.

By 1900, bison were driven to a few hundred individuals. This threw a wrench into the spokes of an ecological system that formed America's agricultural breadbasket. What had been America's greatest renewable food producing resource was made finite and no longer self-sustaining. When European immigrants displaced natural predators like the grizzly, wolf, and indigenous people, it forever tipped the scales out of balance.

Ecosystems have what are called keystone species. Imagine how complex an automobile is. There are hundreds of moving parts that must function together. If you remove just one critical component, it all comes to a halt. Ecosystems have similar vulnerabilities.

A balanced ecosystem has a portfolio of species diversity that allows it to be resilient to fires, floods, droughts, and storms. When European expansion committed mass genocide on native peoples, they removed a key component of America's ecology. When settlers killed the wolves, the ecosystem lost a keystone species that kept bison, sheep, goat, elk, and deer herds strong and their populations in balance. Wolves, coyotes, and foxes kept rabbits, squirrels, and rodent populations in check. When Europeans wiped out the Great Plains carnivores, settlers were faced with a plague of rabbits, mice, and rats.

As settlers moved west inconvenient species were simply moved aside or killed. There were fewer species to keep the grasses and shrubs under control. After European westward expansion, the system no longer had the diversity necessary to regain a naturally balanced ecology. European settlers with guns and plows took over the niches of the wolf and bison.

Beavers build dams that turned streams into ponds that eventually became meadows and forage for grazers. Beaver trapping significantly increased flooding and erosion. Cities, fences, and asphalt cut migration routes, but nothing kept the human population or appetite for consumption in check. What had been a naturally renewable ecosystem for millennia became finite and began to die. Every organism civilization drove to extinction meant humanity must assume that organism's role in the ecosystem. Erosion, forestry, flood control, and wildlife land management became our responsibility.

The following is an example of how the impact of only twenty-nine men and one pregnant teenage indigenous girl impacted an otherwise pristine environment. In his book *Lewis and Clark on the Great Plains—A Natural History*, Paul A. Johnsgard wrote: "In the course of the expedition the group lived off the land, killing and eating almost anything they could. Burroughs compiled a list of game killed in the course of the expedition, largely for human consumption. At minimum, it included 1,001 deer, 35 elk, 227 bison, 62 pronghorns, 113 beaver, 104 geese and brant, 48 shorebirds ('plovers'), 46 grouse, 45 ducks and coots, and 9 turkeys. They also killed 43 grizzly bears, 23 black bears, 18 wolves, and 16 otters" This level of resource exploitation marked the beginning of a century of unrestrained wildlife slaughter in America, ending in the elimination of the bison, elk, gray wolf, and grizzly bear from the Great Plains, and the total extinction of the passenger pigeon, Carolina Parakeet, and Eskimo Curlew. More than one hundred previously unknown species were identified by the expedition. Of those, almost half are now endangered or already extinct.

It is important to note that the Lewis and Clark 'Corps of Discovery' consisted of a contingent of only thirty people for a period of two and a half years. The expedition began with over ten tons of supplies that included dried provisions, 120 gallons of whiskey, trade goods, state of the art weapons, powder and shot, tools, a portable foundry, collapsible iron boat, axes, and other necessaries. Even with all those provisions and well before reaching the Rocky Mountains, they were almost entirely dependent upon resources collected along the trail.

Without the help of Sacajawea, a Lemhi Shoshone woman, most of the Corps would have suffered malnutrition or not have survived at all. On one occasion, she recovered most of the field notes and scientific equipment lost overboard in rough water. After extreme hardship and depleted supplies from struggling over the Rocky Mountains, the team split up with the strongest forging ahead to find food. Sacajawea served as their guide. Suddenly a large band of Shoshone warriors arrived with intent to kill the invaders. Sacajawea stepped forward to parlay only to discover that the war party's leader was her brother. Meriweather Lewis claimed she was as responsible for the expedition's success as any male member of the Corp of Discovery.

As devastating as the Lewis and Clark expedition might seem, their combined environmental footprint was minuscule compared to that of a single US citizen today.

The sustainable dynamic equilibrium between a population and the environment is called the carrying capacity. When a population exceeds the carrying capacity, nature adjusts the population to put things back in balance. Humanity has already exceeded the planet's carrying capacity in at least four areas: extinction rate, deforestation, atmospheric GHGs, and the interruption in the flow of nitrogen and phosphorus through the ecosystem. The latter is vital to the regenerative capacity of living systems.

Civilization's footprint has not remained constant; consumption, pollution, loss of species diversity, mass extinction, and greenhouse gas emissions all continue to accelerate at a terrifying rate. Unregulated growth was manageable in the past. Today it represents the sixth mass extinction and the collapse of those bio-geochemical systems that allow civilization to exist.

Recent reports by the UN, the World Wildlife Fund, and the Nature Conservancy state that up to one million species are threatened with extinction by 2100. As many as 150 species go extinct every day. Today's sixth mass extinction is directly attributable to civilizations impact on the environment. To put this in perspective, the previous record for the most rapid mass extinction took place just over sixty-five million years ago with the extinction of the dinosaurs. It probably took over twenty thousand years to eliminate the last pockets of dinosaurs. The sixth extinction event that we are responsible for is taking place at a rate one hundred times faster, with nearly 50% of all large land animal species becoming extinct in the last 200 years. The extinction rate in the last fifty years is happening thousands of times faster.

Humanity's footprint increases at a rate far greater than increases in the population. In the 20th century, the global population exploded from 1.6 billion to 6.1 billion. While the population grew four-fold, global emissions increased twelve-fold or three times faster. The population surpassed eight billion in 2022. We can expect a population well over nine billion by 2050. The exponential trajectory of civilization's footprint is likely to increase more than twelve times that of population growth because developing countries will acquire developed nation appetites. That means civilization will consume more in the next twenty-three years than all of humanity since the industrial revolution. More energy will be required over the next twenty-three years than was used since the invention of the steam engine. This mathematical reality should make it clear why humanity must transition to more dependable and sustainable sources of energy to take civilization into the twenty-second century and beyond. Policy makers must system-think in terms of exponential rather than linear growth.

Energy forecasts have special importance to the economic growth of developing countries. No nation should shirk their responsibility to other nations. There is an urgent need to ensure that developing nations do not commit their development to fossil fuels. Once a nation starts down the fossil fuel path, it becomes increasingly difficult to transition to sustainable sources when scarcity and rising costs compel them to. It falls to wealthy nations to guide responsible development with financial and technological assistance. If wealthy nations fail in this, most future emissions will soon come from developing countries. If that happens, whatever measures wealthy nations make to reduce their own emissions will have little effect in curbing global GHG totals.

Prior to China's economic boom, the US remains the largest total contributor of GHG emissions. China is the current number one emitter and the second largest economy in the world. By 2020, China's enormous industrial growth and population had vastly increased their national footprint, though the per-capita footprint remained relatively small. For that reason, China is still defined as a developing country. Individual prosperity in China is rapidly catching up. China's appetite for western consumption is spreading. Soon China's per-capita footprint will match that of developed nations. China's wealth and economic status now means it must step up to the plate and share the cost and responsibility to limit emissions 50% by 2030. If it doesn't, nothing the rest of the world does will prevent warming past 5.4°F (3°C) or more.

India is the world's fourth greatest emitter of greenhouse gases, but their per-capita environmental footprint is also small. As individual prosperity improves on the subcontinent, India's collective per-capita footprint will grow much faster. A similar argument may be made for Indonesia, the African nations, and Brazil. Even developing nations must recognize their global responsibility to develop in more sustainable ways.

Is there another reason these nations should reduce their emissions? A few years ago, President Clinton was asked why China was committing hundreds of billions of dollars to totally transition to sustainable energy by 2060. His answer could not have been more to the point. He said, "They can't breathe."

It is a sad fact that those least responsible for causing global warming are the ones that are suffering the most from it. When people can no longer find food or water, or maintain security, social systems fracture, they tend to get rowdy, march in the streets, migrate, or start shooting. All food, water, and shelter come from the environment. Wars may be about race, religion, or equity, yet all of those motives evolve from the lack of environmental security. In addition to halting population growth, economic policies must reflect the value of a sustainable environment. It is not only the direction of change that is necessary, but change must happen at an unprecedented pace.

Chapter 14
Gross Domestic Deception

We are in the Anthropocene. Human psychology must now be given a higher priority in the calculus for sustainability. The mechanism to bring about change in civilization is politics. That is a problem because most Americans don't believe politics is working as well as it needs to. The lack of good governance and honest politicians became obvious when the US was faced with COVID.

Sinclair Lewis presciently wrote about an imagined fascist takeover of democracy in America. That theme carried throughout his writing in *Main Street, Elmer Gantry, Dodsworth*, and *Babbit*. It was finally put into clearer focus with *It Can't Happen Here* (1935). Mr. Lewis' premise was often summarized this way: "When fascism comes to America it will be wrapped in the flag and carrying a cross." Lewis' protagonist, Buzz Windrip appealed to voters with a mix of crass language and nativist ideology. Once Mr. Windrip was elected, he energized his working-class base against immigrants and the liberal press. There was the suggestion of ubiquitous and unseen forces threatening local or national patriotic values. A similar theme was abundantly evident throughout the Trump campaign and far right elements of his administration. He claimed that immigrants were taking American jobs Lurking in the shadows of liberal politics was a Jewish cabal. Not true of course but useful tropes for the Republican agenda.

It's an old formula. Hitler used it in Germany. Mussolini used it in Italy. Lenin used it in Russia. There are variations on this theme in the archives of every despot and dictator in history. The formula lasts because it works, and it works because there are innate patterns in the genome of human social psychology that have remained over millennia that allow it to work. Humans have always had difficulty distinguishing between belief and reality. Some of the strongest beliefs have little or no evidence to support them.

In 2017, Peter McIndoe wondered how outlandish a conspiracy theory would need to be before it was rejected out of hand. To test American gullibility, he came up with something totally irrational. What if someone claimed that 'birds aren't real'. They're actually drones designed to spy on people. Would anyone believe such nonsense?

Peter drove a van professionally painted to look official. His website was designed to look like a whistleblower was uncovering a legitimate fact. His blog soon had hundreds of thousands of followers. Soon independent spin-off groups began organizing to spot and track drone birds. In a dozen locations around the US. Billboards sprang up in Pittsburgh, Memphis, and Los Angeles boldly declaring 'Birds Aren't Real'. They charge their batteries by sitting on powerlines. They fly into windows and poop on cars to mark suspects for follow up investigations. There were even blueprints to show how they worked.

Mr. McIndoe wanted to point out how ridiculous conspiracies and misinformation had become. He hoped even the most rudimentary reflection would bring people to their senses. Would people believe the nonsense without thinking? No one seemed to be getting the punchline. When Mr. McIndoe announced it had been an experimental joke, some refused to believe him and started another conspiracy that he had been brainwashed to recant. Some became hostile and continued to believe the yarn despite Mr. McIndoe's confession.

There doesn't seem to be any limit to what people will believe. Psychologists tell us that the human brain absorbs information without judgement. It's just information. It takes training and discipline to learn how to separate fact from fiction. Experience and education are necessary for us to learn how to segregate and prioritize information. However, even training may not be sufficient. Social pressure can still motivate us to believe a good yarn more than we believe reality. Most people are willing to accept 'alternative facts' when presented with a good story by an accomplished pitchman. QANON is an example where conspiracy theories are accepted as fact and become a powerful influence on society despite having little support from credible evidence.

As social animals, we are compelled to trust each other. Sociopaths will use that trust to deceive us. Extreme right politicians express outrage when compared with pre-war German propaganda. That doesn't negate the fact that their behavior closely matches the same fascist ideologies and methods. A good example of the manipulation of trust can be found in the archives of one of the best pitchmen of all time, P.T. Barnum.

The 'Cardiff Giant' was carved from stone and passed off as a prehistoric giant. Experts soon proclaimed it to be a hoax, but much of the public continued to believe in its authenticity. In 1869, a man named Hannum purchased the fraudulent statue. He charged admission, continuing to claim it was the real thing. Circus mogul P.T. Barnum tried to buy it, but Hannum refused. Undaunted, Mr. Barnum created his own replica, claiming his giant was the real item, and Hannum's original carving was the fake.

When Hannum learned of this, he spoke those now famous words, "There's a sucker born every minute," referring to those gullible customers who paid Barnum to see his fake.

A carnival barker, or one occupying the Bully Pulpit in the White House can also use trust to dupe the public into buying a ticket. Mermen, Atlanteans, and the conspiracy theory that the 2020 election was stolen are just a few more examples where people can be led to accept a good story by a gifted pitchman. To a significant portion of the public, evidence is irrelevant. Politicians have known and used this characteristic of humanity for millennia. Abraham Lincoln said, "You can fool some of the people all of the time. You can fool all of the people some of the time, but you can't fool all of the people all of the time." Despots are well known to pitch alternative facts to further their own purpose.

People tend to believe false information when 'alternative facts' are portrayed as the truth by someone they trust. Blatant sinners might preach fundamentalism and cite the word of God. In their defense, they will accuse others of heresy and worshippers of false gods. The greedy will feign compassion and philanthropy, hiding the same miserly *noblesse oblige* of feudal aristocracies. Meanwhile usurpers plot to advance their own self-interests with deceit and lies. They will cry foul and accuse those who might oppose them of spreading fake news. The accusers become the accused. When lies go free, there is no truth.

Another sleight of hand tactic to advance a false ideology is to distract the public's attention with outrageous hyperbole, and a Gish gallop of rapid-fire BS to overwhelm the opposition.

If we recognize these patterns, the fog begins to clear. For over half a century, we have been told America was the wealthiest nation in the history of the world. National wealth and public prosperity are two entirely different things. Wealth in America really consists of a handful of the world's richest people, a shrinking middle class, and the threat of instant poverty for the rest.

Imagine two identical rooms. Let's put ten people of average income in one room. We'll call that room 'A', the Scandinavian room. Now let's put another ten people in room 'B'. We'll call room 'B', the American room. We'll set the average personal annual income in both rooms at $40,000. The only exception will be one person in the American room ('B'). The total wealth of room 'A' is ten times $40,000. That gives room ('A') wealth a total of $400,000.

Now let's look at the American room ('B'). Nine citizens in the American room also make $40,000 a year. Their combined wealth is nine times $40,000, for a total of $360,000. The remaining person in room 'B' is Elon Musk. In 2021, USA Today reported Musk's net worth was around $214,000,000,000.00 give or take a few billion. That means room 'B' is now worth $214,000,360,000.00. Room 'B' is the richest in total dollars, but can room 'B' really be considered the wealthiest room (nation) when nine out of ten make the same as those in room 'A'?

America's national wealth may be the highest, but the per capita income of the average citizen is less than those living in Qatar, Macau, Luxembourg, Singapore, Ireland, Brunei, United Arab Emirates, Kuwait, Switzerland, or Norway. (International Monetary Fund 2020).

This example looks even worse when we flesh out what the average citizen in America gets for their dollar compared to what other countries get. Does GDP accurately measure the wealth of a nation if over 40% of the wealth is in the hands of the top 1%? Is real wealth measured in productivity and total national dollars or should it be measured by something else? Many economists and sociologists would argue that no amount of productivity or financial wealth is meaningful if it doesn't translate to improving the security of every day public needs and wellbeing.

The numbers tell a damning story. America spent $6.4 trillion on war in the Middle East, but balks at spending half that amount on infrastructure, education, science, technology, climate change, health care, transportation, agriculture, decarbonization, job training, social services, high speed web, or cybersecurity.

John Maynard Keynes said, "The difficulty lies not in the new ideas but in escaping from the old ones." Let's start with an obvious component of modern infrastructure, health security. Most people would agree that it is better to be alive and healthy than near death and rich. Most would also agree that you can't build a civilization without a healthy work force. Yet, America is the only industrialized nation without universal health care. When the COVID pandemic hit, forty million had no health insurance at all. They were told to go to the doctor but couldn't afford to. An untold number that got sick or died at home were unreported. That masked the actual number of cases spreading around the country.

In February 2020, nearly half of Americans had less than four hundred dollars in their savings to live on during months of quarantine, health care and shelter-in-place. One in ten Americans became unemployed in just eight weeks. There was no guarantee they would regain employment after the pandemic subsided. America's national health system was totally inadequate to the task of dealing with a pandemic emergency. It was necessary to ask assistance from other nations to obtain basic hospital personal protection supplies (PPE). Those supplies weren't for the sick, they were for emergency responders, doctors, and nurses that care for the sick.

The blow to the US economy was suddenly measured in trillions of dollars. Trillions more would be necessary to pull the economy back to anything resembling normalcy. Basically, COVID shot America's health security all to hell with a single blow. Environmental scientists and health specialists warned that COVID was only round one as climate change continued unabated. A warmer planet incubates new and more virulent diseases.

There are many ideas of what personal wellbeing means. The economic view includes things like income, housing, and purchasing power. The more social perspective places a higher value on quality of life, education, retirement, social stability, confidence in the future, and perhaps most of all, general happiness. How does American wellbeing compare with other developed nations from a purely pragmatic viewpoint? Basically, it sucks!

A significant indicator of wellbeing and social equity is education. In the US there has been no improvement in reading since 2000, no improvement in math since 2003, and no improvement in science since 2006. The US is falling farther behind every year. According to the National Assessment for Educational Progress, math and reading scores for fourth and eighth grade students has been dropping rather than improving. In 2018, the US still ranked thirtieth in math, eighth in reading, and fifteenth in science compared with the sixty-five nations surveyed. How can America compete with those scores?

The educational performance gap between economic classes in the US has widened. This reflects a political divergence that inhibits universal access to quality education. The cost of higher education is already moving out of reach for millions of qualified American students.

In 2020, American students attending an average four-year college had a student loan debt of roughly $40,000 upon graduation. The average student attending an ivy-league university will owe more than $100,000.00 at graduation. Basically, they have the debt of a home mortgage without the home. This impacts the overall economy in several ways. Student debt suffocates the energy, innovation, and brain power of America's youth, further reducing US competitiveness in the global economy.

Despite these official government figures, neo-conservative leaders continue to claim that America is the richest and most powerful nation in the history of the world. They may be right if they only look in the mirrored halls of Congress (room 'B'), and the major contributors to their campaign funds. They may be right if they only look at Wall Street, and the GDP, but those are not the best indicators of national wellbeing. The gap between the mega-rich and working class is stunning and becoming wider by the second. The average CEO to wage earner salary ratio in 1961 was 21/1. The average CEO to wage earner salary ratio in 2020 was 351/1. It should become clear what sector of society is profiting the most from unregulated economics and far-right politics.

How does US wellbeing for the average citizen compare with other developed nations? America ranks twentieth in the world for infant mortality, twenty-sixth for life expectancy and fifty-fourth for maternal mortality behind Turkey, Uruguay, and Tajikistan. The US rate of maternal mortality has nearly doubled since 1970 while the global aggregate rate has continued to improve. That's right, a mother giving birth in the US today is more likely to die than in 1970.

On June 24, 2022, the US Supreme Court overturned Roe v. Wade, reversing a woman's constitutional right to abortion and reproductive choice that had stood for two generations. That decision placed further burdens on women's health and control over their own bodies.

The US ranks tenth for gross per capita income. This is made worse because the average US citizen carries one of the largest personal debt loads in the world. Home ownership in America ranks forty-fourth based upon 2018 data. America ranks twenty-seventh in the world for overall education and health, a marked decrease from sixth in 1990, largely due to decreased funding for public education grades k-12 and inflation in the cost of health care.

A Business Insider article published on September 18th, 2018, stated that, "By improving its education and healthcare policies, the US could see faster economic growth." Political conservatives promoted defunding quality public education under the idea that private and charter schools offered greater choice, but like health care, the ability to choose higher quality education only applies to those who can afford it. Quality education in the best charter or private schools is expensive. That creates a bias in favor of the wealthy and evolves into an aristocracy class. The effect is to segregate economic and ethnic portions of the population and turn the civil rights clock back half a century.

Representative democracy requires an educated electorate. Despite the advice of economists, educators and social scientists, America has continued to defund the very institutions that make it competitive in a highly competitive world. Americans now work longer hours and experience more stress related health issues than nearly all other industrialized nations.

When justice, equality, a collapsing infrastructure and education is ignored, America becomes less resilient than other developed nations. The government as well as the average citizen is less able to adapt to setbacks. America's wealth has failed to correspond to general wellbeing. Rising national wealth has had an inverse relationship to personal happiness, resilience, competitiveness, education, and health for the general population (Robert Reich, January, 2020).

There is a more subjective perspective on civilization. At its core, it is what human life is all about. Is the goal continual struggle and suffering or is it the pursuit of greater happiness and wellbeing? I'm reminded of a scene in a Star Trek movie where Mr. Spock's father asked young Spock, "How do you feel?" The young Spock was puzzled. He didn't see the relevance. His father's point, however, was that it may be all that was truly relevant. How do we feel living in America? Are we content with our lives? Do we feel our future is secure?

Are we enjoying life and having ample opportunity to just have fun and enjoy our family and friends? The 2019 World Happiness Report stated that "The U.S is the unhappiest it has ever been and is sinking lower each year."

According to another 2019 report, the four happiest countries in the world were Denmark, Switzerland, Norway, and Finland. These countries are social democracies with free, open, and secure elections. In 2018, the US dropped to nineteenth place from fourteenth in 2017, while Finland ranked as the happiest nation on Earth. According to the 2019 World Happiness Report, "All of the top countries tend to have high values in six key variables found to support wellbeing. Those key variables are income, healthy life expectancy, social support, freedom, trust, and generosity. Among the top countries, differences are small enough that year-to-year changes in the rankings are to be expected." The 2019 *World Happiness Report* also looked at immigration. There was a strong correlation between the top ten happiest countries, and the happiness of immigrants in those countries. Immigrant scores were nearly identical to native citizens in all six key variables. There was a significant correlation between social equity and success in supporting the six key components of happiness.

The most combative political debates in the US continue to be about the type of government rather than good governance and striving to improve general wellbeing. One of the strongest indicators of good governance is the status of women. The UN has firmly established that the status of women is the single strongest indicator of a stable, healthy, and sustainable nation. "Empowering women in the economy and closing gender gaps in the world of work are key to achieving the 2030 *Agenda for Sustainable Development*."

For example, according to a report published in Social Europe by Lorenzo Fioramonti (et.al.), June 1, 2020, those nations with a female leader had six times fewer deaths from COVID than those states with male leaders. That same year the UN Security Council report *Women and Peace and Security* stated that women have demonstrated the dominant role in providing front-line health care during the COVID epidemic. The report noted that nations that failed to provide equity for the rights of women had higher rates of economic instability, poorer health care in general, and a greater tendency for internal violence, and international conflict risk.

Americans have had their noses to the grindstone for so long they have developed a kind of cultural cynicism. Some readers may wonder what relevance 'happiness' has to do with anything. Some may experience guilt in admitting they even strive for happiness. It might be important to take a moment to think about what governance and the economy is all about, if it isn't to improve wellbeing and happiness. This relates to climate change because unhappy people are less able to solve complex problems.

Mark Twain said, "It ain't what you don't know that gets you into trouble. It's what you know for sure that just ain't so." Most people believe modern civilization has made a great contribution to the wellbeing of humanity. Defenders of the modern, mostly urban lifestyle, cite higher incomes as evidence. Incomes are higher, but has the pursuit of wealth overshadowed the pursuit of greater wellbeing? Professor David C. Korten, author of the international bestseller *When Corporations Rule the World* said, "We will prosper in the pursuit of life, or we will perish in the pursuit of money. The choice is ours."

Most Americans are convinced that civilization is adapting to human needs, but is it possible that the opposite is true? There are significant signals that people do not function well in the crowded, rapidly evolving complexities of the 21st century. James Suzman found after decades of research on hunter-gatherer tribes that neolithic populations worked an average of only fifteen hours a week. In 2020, the average full-time American male worked 41 hours per week. Our neolithic human genome was well adapted to function in small bands of a few dozen to perhaps a few hundred. But behavior traits that were an adaptive benefit over the past 300,000 years, may not have served humanity as well in the last 10,000 years or the last 200. As the 21st century population increases, the social stresses and environmental impact increases far out of proportion. It is becoming more apparent that we are genetically maladapted to live in this world of our own creation.

By midcentury, nearly two thirds of the global population will live in dense urban environments. People living today work longer hours with more social stress than they did ten thousand years ago. They spend less time participating in family activities, raising children, or participating in social bonding activities. People living in cities suffer more stress related illnesses. Urban populations are more likely to become neurotic, anxious, and hostile than so called primitive societies. Excessive stress reduces resilience and adaptability.

It wasn't an opposable thumb or standing upright that brought *Homo sapiens* to global dominion. Many organisms had special capabilities that exceeded ours. We didn't have keener hearing or sharper sight. Many species were faster or stronger. There were several times when the environment was so hostile that the human branch of the family tree was nearly pruned to extinction. (Seventy-two thousand years ago, the Toba super volcano in Sumatra almost immediately dropped the global temperature as much as 27°F (15°C) for as long as one thousand years. Some estimates suggest the total global human population was reduced to as few as 1000 breeding pairs.) It is unlikely that survival resulted from rugged individualism. Why did humans succeed when others failed?

Social organisms find strength in community. Peck-order, conformity, and obedience to authority are three innate traits that contribute to social cohesion. People quickly recognize what position they play on the team. They learn the rules of the game and pay close attention to the captain to tell them the plan. Human societies are built on the experiences of those who came before. Collective experiences were passed on through stories, expanding the human playbook with every new generation. The benefits of sharing work and assigning responsibilities were pivotal to survival and the formation of civilization. As knowledge and experience grew, old repetitive problems were eliminated while new challenges took their place. Humans are strong team players, and as team players, most play by the rules. Those small population bonds worked well in hunter-gatherer bands, but urbanization may be turning those traits against us. As the climate moderated during the Holocene, small community settlements eventually coalesced into more complex urban developments. Civilization is often defined by highly diverse and specialized skills, social stratification, some form of government, symbolic communication, and a system of numeration to track and regulate trade. Institutions evolved to monitor and regulate the economy and society. These were the hallmarks in the evolution of early civilizations.

Intelligence wasn't the only thing that brought humanity to global dominance. Intelligence is a transitory thing unless experience can be shared and passed on. Social cohesion in civilization demands a sense of duty and self-sacrifice to the group. The most successful leaders are able to pull people together to tackle difficult tasks. The importance of good leadership is even more important today because the consequences are more dire.

In October of 1957, the US suddenly realized it had fallen far behind the USSR in rocket technology and was vulnerable to intercontinental missile attack. Global security depended upon parity between the two greatest nuclear powers. Small gains wouldn't be enough. America needed massive strides to catch up and surpass the Russians. The scope of the mission was almost unimaginable. Something decisive had to be done to re-establish the balance of power. The strength to accomplish that task was not in the government, but in the combined will, experience, and skill of the American people. A leader would have to unite the people with common purpose.

On January 20th, 1961, President John F. Kennedy said, "Ask not what your country can do for you, ask what you can do for your country." On May 25, 1961, President Kennedy focused America's people-power on two objectives. The first goal was to develop new rocket technology. The second was to use that technology to put a man on the moon by the end of the decade. Both tasks seemed impossible. But Kennedy realized the immense power available when Americans pulled together. He believed his role was to set the goal and the government's role was to provide continued focus and support.

With the task clearly set before them, the American people went to work. On February 20th, 1962, John Glenn orbited the Earth three times in a space capsule named Friendship 7. The *unimaginable* took less than nine months. On July 20th, 1969 Neil Armstrong and Edwin 'Buzz' Aldrin became the first humans to land and walk on the moon. The *impossible* only took seven years and nine months. If we can walk on the moon in seven years and nine months, we can stop belching GHGs into the atmosphere in the same length of time.

Today's leaders must set a clear goal and command the same sense of public service and self-sacrifice Kennedy asked of the American people. America plays a critical role in the world just as it did during the Cold War. American leadership must set the goal of addressing climate change and then guide and support the global community to achieve that goal. The climate crisis demands that every nation function as part of a single team.

The United Nations is the only venue with a successful international track record on climate change. The 193 members of the UN have already done most of the strategic planning. American leadership would help restore early progress on climate change and the seventeen Sustainable Development Goals (SDGs).

The World Resources Institute (WRI) reported that greenhouse gas emissions grew 50% between 1990 and 2018. It doesn't take a rocket scientist to see that trend is a swiftly growing threat. All nations must work together to develop the technology, build the infrastructure, and reduce net GHGs 50% by

2030 and then reach net zero by 2040. That still won't be enough because remaining emissions will continue to warm the planet for hundreds of years.

Technology must be developed to pull GHGs out of the atmosphere to remove those earlier emissions. Transitioning from fossil fuels may seem unimaginable. Achieving net zero by 2040 may seem impossible, but President Kennedy proved that engineering the *unimaginable*, and achieving the *impossible* is what America does best.

US politics is sloppy, dirty, and cantankerous. No system of government is perfect, but it is not the system of government but the effectiveness of good governance that will determine if America and the world will be successful in building a sustainable future.

The United Nations measures good governance by eight factors: participation, rule of law, transparency, responsiveness, consensus orientation, equity and inclusiveness, effectiveness and efficiency, and accountability. A civilization fails when it no longer maintains the three Cs of community, communication, and cooperation.

We see a rise in authoritarianism and a decline in American democratic values today. There are powerful forces working against domestic and international cooperation. Enter the mega-corporation, the seditionist, and demagogue usurper. Some of those exhibiting the greatest financial success are the most predatory. Their relationship to civilization is parasitic and sociopathic. They do not measure their progress in terms of human wellbeing. The very reason of their existence is self-interest.

Giant monopolies are fundamentally sociopathic fungi whose tendrils weave into the very marrow of the human society, the economy, and the environment. Corporations are a human construct and perhaps the ultimate expression of the hunter-gatherer, apex predator pathology. Corporations may qualify as legal persons, but they do not exhibit normal human social behavior. Avarice is their motivation. What they hunt and accumulate is theirs alone and humanity is their prey. Corporate sociopathic behavior is often a reflection of their leadership. Recent studies reveal that approximately 20% of corporate executive officers exhibit sociopathic tendencies (Forbes, 2019; Business Insider, 2017). Those CEOs are highly skilled at social manipulation. Their personal lives are antisocial and manipulative, unless they determine that it enhances their sense of power and importance. They tend to be thrill seekers. They exhibit behavior that may place others at risk. They like to pass blame for their mistakes onto others. They tend to be vengeful. People who mess with them will regret it.

A few will carry anti-social behavior to psychopathic extremes. The psychopath is less likely to suffer remorse or guilt than the sociopath. The psychopath will lie regardless of the consequences so long as it defends or advances their personal objectives. Empathy does not exist in their pathology. Those who disagree are quickly cast aside as the enemy. The sociopath will use 'alternative facts' without evidence to defend an indefensible position, while the psychopath will unshakably believe their own alternate reality, regardless of overwhelming evidence.

Sociopathic behavior influences every facet of civilization. It contributes to the divisions we see in politics, society and the erosion of good governance. It compromises democratic representation and diverts the benefits of economic growth to an elitist minority. Self-interest is what motivates the corporate mindset. Corporate resistance to paying an equal share of taxes is an example of a sociopathic mindset. It insists on equal voice but not equal responsibility to the human community that pays for it or the environment that feeds it. In media and political debates, America hears complaints that taxes are too high. Corporations and the wealthy claim that higher taxes threaten economic growth, but the numbers simply don't support that claim.

Billionaire Warren Buffett once pointed out that he pays a lower tax rate than his secretary. Secretaries can't deduct business luncheons on their five hundred million dollar, four hundred seventeen-foot-long private yachts like the one owned by Jeff Bezos. If that doesn't take the frosting off your cookies, think about this. Among the multi-billionaire set, Mr. Bezos' yacht is only medium sized.

According to ProPublica, Elon Musk (the richest man in the world) paid '0' tax in 2018. One in five of the fortune five hundred corporations in America also paid no tax in 2018. The other four fifths paid a substantially lower tax rate than their average employee. When corporations pay a smaller proportion of the cost of civilization, the burden is passed on to the taxpayer and future generations by the national debt. When the national debt grows, the interest on that debt is also passed to the red ink column of the ledger. In 2021, the interest alone was nearly four hundred billion dollars. Meanwhile, mega-corporation and bank executives have a party in the Caymans. In their isolated, insulated, and gated communities, they don't see or feel the impact of climate change as the planet continues to bake and the general public pays the check.

An unregulated consumer economy and the extreme-right agenda is only giving lip service to social wellbeing. The numbers don't lie. The wealth is going to the 1% all-powerful elite. The GOP claims to be the party of law-and-order. They claim to promote justice, peace and security, but at their rallies they promote chaos, violence and divisiveness. The GOP broadcasts QANON conspiracies about secret subversive forces that undermine America's Christian moral foundation.

Meanwhile, Trump appointed Supreme Court Justices overturn long standing constitutional human rights. In the states, Republican minions seek to control the right to vote. Deregulation and voting restrictions focus power and wealth away from public wellbeing to corporations and increased power to an illegitimate authoritarian leader. That is the very definition of fascism. Despite all the warnings, the decline of American democracy and growth of fascism is taking place in plain view.

Plato said, "The price of apathy toward public affairs is to be ruled by evil men." The most vocal complaints about high taxes are coming from the white-collar executives and stockholders who make their money by investing in giant sociopathic corporations. Wealthy individuals must pay taxes on the incomes they make from their salaries, stock holdings, and bonuses. Imagine you're the CEO of a mid-sized corporation, with a paltry salary of ten million dollars a year. You receive an annual bonus of an additional couple million and a few more in stocks. If you were an average taxpayer, you would have to write a big check to the IRS, but as a wealthy CEO you might choose another option.

You might point out to your Senator or Representative that it might not be possible to make a substantial contribution to their campaign without some tax relief. You might meet with that official in Bermuda, or some other out-of-the-way luxury resort. On the golf course, you might talk a little business, so you might be able to deduct that. You could travel there in your private jet and write a memo on the way. You could deduct some portion of that as a business expense. You might even run for office yourself. You might run for President or become a Supreme Court Justice. That way you could protect your portfolio by voting to reduce regulatory oversight and thereby increase corporate profits. You could become a corporation yourself and brand the use of your name. As the face of that brand, you might deduct nearly everything you do without providing any real service at all. Of course, taking money while providing nothing substantial might be viewed by some as fraud. The law might look the other way if you were able to influence enough politicians or stack the courts in your favor. It worked for Al Capone…for a while. Today it works because public cynicism expects big business, the wealthy and government officials to lie, cheat, and steal.

The relationship between the state and corporations is a complex one. Ambrose Pierce's 'Devil's Dictionary' defined a corporation as 'an ingenious device for obtaining profit without individual responsibility'. He went on to say, "It is a legal construct, a charter granted by the state to a group of investors to gather private funds for a specific purpose. Originally, charters were granted in the service of a public purpose and could be revoked if this were not fulfilled. Over the past 400 years, corporations have conquered territory and brought in resources for the state, gaining in power and privilege. History shows a repetitive cycle of corporations over reaching, causing such social turmoil that the state is forced to reign them back in through regulation."

Corporate charters have played an enormous role in global history. A charter was often granted by a nation to improve cash flow, to further foreign conquest, or pay for a militia, mercenaries, or even privateers (mercenary pirates). Corporations were sometimes granted the authority to establish courts, levy taxes, form monopolies, maintain armies, and even negotiate treaties. The corporation became the British Empire's tool to plunder the resources discovered in the Americas, Africa, East Indies, and Far East. At its height, the British East India Company ruled over a fifth of Earth's population with a private army of a quarter of a million. The Hudson Bay Company was loved and hated for similar reasons.

Early American small businesses resented the heavy boot of powerful corporations. The American colonies wanted to manufacture native materials for their own consumption. Corporate interests and the British Monarchy demanded that raw materials be shipped to England, manufactured there, then shipped back to America and sold at exorbitant prices. British corporate control of trade was strangling the American economy, ultimately pushing it toward declaring independence.

Much of the inability of America to address climate change comes from fossil fuel corporations protecting their profit margins. To them, profits trump (not a pun) human or environmental wellbeing. Corporate manipulation of economic policy can be seen in the pharmaceutical industry as well. A few pharmaceutical giants have manipulated legislation so that they set the price on medicine, not the competitive marketplace that unregulated free-marketers claim. The average citizen is not allowed to seek and pay the lowest price.

A democracy cannot affectively address climate change, or any crisis when monopolies have greater political influence than the public. A collapsing environment is invisible if the government is ignorant, incompetent, corrupt, or all three. In recent decades, the US Supreme Court has made achieving good governance far more difficult by favoring corporate persons over biological persons. This was a major coup for the Republican Party and monopolistic mega-corporations. It offered greater corporate political representation than was available to the average voter.

In 2010, the Supreme Court of the United States reversed long standing campaign finance restrictions. In their Citizens United vs The Federal Elections Commission decision, the US Supreme Court allowed corporations to contribute funds to elections under the naïve assumption that it would not result in increased corruption or bias the election process. That decision allowed unlimited funds to be contributed anonymously to serve corporate interests. The five to four court decision was justified under the Freedom of Speech clause in the First Amendment of the US Constitution. It was also defended under the right of equal representation granted by the Fourteenth Amendment. The court ignored two key differences between humans and corporations. Corporations don't die and you can't put a corporation in jail. The Citizens United case did precisely what the court said would not happen. Big Money now had leverage on politics and their message. Big Money flowed through think tank organizations like the Heritage Foundation to shills and mass media like Fox News. Truth in journalism was no longer sacred. Remember, a sociopath is not bothered by the truth, so long as it aligns with a self-serving purpose.

Vast sums of special interest campaign money quickly began to overshadow the voice of the people. In the 2018 election cycle, only one hundred donors contributed 78% of all Super Pac spending on the campaign. In 2021–2022, Koch Industries pitched in $8,895,231.00, Occidental Petroleum added $5,562,547.00, Chevron gave another $5,008,598.00, and continued to in 2022 according to OpenSecrets.org.

Corporate and special interest money threatens the health and security of all life by hindering action to decarbonize the global economy and address climate change. How do we solve the problem of good governance when money and self-aggrandizing demigods exercise a greater voice in the media, politics, and the economy than the electorate? In September 2022, President Biden called a spade a spade by identified a growing fascism in America, fully franchised by the Republican Party.

Recently, I was a guest presenter for Planet Helpers, an environment club at the Hazel Wolf STEM (Science, Technology, Engineering and Math) School in Seattle, Washington. I presented the problem of climate change inaction to the class. I asked what they thought was causing America's delay in acting on climate change. They told me that Big Oil and corporate special interests were spreading lies about climate science. I asked if they could think of examples. A very bright eighth grade girl replied, "When corporations have greater influence in drafting laws than the electorate, you have an oligarchy. When an authoritarian leader promotes violent populist nationalism and racism as patriotism, you have fascism." I don't think I have ever found either of those terms defined so succinctly. I wondered why an eighth-grade girl could see the problem while so many adults could not.

In the past several decades, we have seen the rise of vast, monolithic, multi-national corporations. A few, including half a dozen fossil fuel corporations, became so powerful that they gained significant influence over international security and the entire global economy. Royal Dutch Shell, Exxon, Chevron, BP, Koch Industries, and others have become so wealthy, and multi-national in scope, they exceed the GDP of some of the nations they do business with. US progressive and conservative administrations have continued to generously subsidize these corporations. In return, these corporations subsidize favored politicians with campaign donations in their effort to maintain the fossil fuel energy status-quo. The economic benefits of transitioning to sustainable energy and decarbonizing the economy have never been clearer or more thoroughly documented. Meaningful action remains stalemated by a deliberate campaign of disinformation by those same corporations and their supporters.

The responsibility to rebuild good governance ultimately falls to the people. However, the American electorate faces a voting challenge as it tries to restore the moral virtue of their elected representatives. Some states and politicians actively work to restrict voting rights. When fewer people vote, there is a better chance that a motivated minority can carry an election. Instead of working to eliminate voter restrictions and disinformation Senator Ted Cruz (R-TX), Representative Marjorie Taylor Greene, (R-GA), Representative Jim Jordan (R-OH), Minority Speaker of the House Keven McCarthy (R-CA), Governor Ron DeSantis (R-FL), Congressman Mat Gaetz (R-FL), and Governor Greg Abbott (R-TX), to name only a few, actively promote voting restrictions.

Joseph Goebbels used Gish gallop propaganda tactics to achieve Nazi objectives. More recently, there has been a growing tendency for politicians to use similar tactics to promote self-interest objectives. Lies, slander, and deception have become common. Corrupt politicians mask those deceptions with a bewildering cacophony of unsubstantiated BS. The next time you watch a debate or interview with one of the afore mentioned politicians, notice how they interrupt and continue to dominate the microphone. That is Gish galop.

The US government became more authoritarian during recent conservative administrations. The balance of power shifted toward the Presidency and Supreme Court. The more extreme wing of the Republican Party used a tactic of accusing opponents of being fascists, or socialists (implying communist). When the mainstream media cried foul, Republicans switched tactics and accused the media of reporting fake news. A deluge of scandalous Gish gallop accusations was used with little regard to their historical accuracy or evidence. Accusations were hurled back and forth until the public didn't know whom to believe. Much of the media focused on the personalities and vitriol, rather than the who, what, where, when, why, and how of the news.

In the ensuing chaos, the contagion of sociopathic corporate influence and authoritarianism continued to grow. Few would have the courage to cry foul or name names. If anyone's hands were dirty, the usual GOP response was to pound the podium and shout that the opponent's hands were dirtier. Name calling and political chaos would get much worse. Big Oil and Big Money continue to fund the takeover of democracy. The GOP is still Trump's party while a masterclass in racketeering continues unabated. Could the facts come to the surface, and if it did, could justice be served with GOP packed courts willing to legislate from the bench?

Chapter 15
Precautionary Principle and Security

The Anthropocene poses both the greatest promise and the greatest danger to the security of all life on Earth, yet there are brighter angels of humanity everywhere. Child prodigy Jackson Oswalt built a fusion reactor in his parents' garage when he was twelve. Emily Bear wrote her first song at three, began publishing her music at four, performed her first concert at five, and played at the White House at six. Alma Deutcher wrote an opera at eight and had it performed in Vienna at age eleven. Children are doing genetic research in high school science labs and designing and building solar powered cars that compete in races across Australia at over sixty miles per hour. Children are the hope for the future, but we must give them the tools and a secure environment for them to succeed. If we understand the dangers they face, we can begin to *mitigate* threats, build *resilience* to new threats, and increase their ability to *adapt* to change. Improvements in science have only added clarity to that understanding, but action has been far too slow because science has been ignored or intentionally hidden.

Alisa Singer did the artwork for the cover of *the 6th Assessment Repor*t by the IPCC. She summarized this generation's obligation to children and their future when she titled her work 'A Borrowed Planet: Inherited from our ancestors, on loan from our children'. The immediate task, however, belongs to adults to give children hope. As the adults, it is our responsibility and no one else's. What course of action should we take? Should we sit back and hope for the best or should we take action to prevent the worst? This choice forms the core of what policy makers call the precautionary principle.

A report appeared in the September 2001 issue of *Environmental Health Perspectives* that identified four central components for environmental decision making (D. Kriebel, et al): "Take preventive action in the face of uncertainty; shift the burden of proof to proponents actually working in the area, field or activity; explore a wide range of alternatives to possible harmful actions; and increase public participation in decision making." The best-case scenario of climate change is still an existential threat to global security by magnifying the probability of wars, disease, and famine.

Environmental security has many faces. We lock our doors to secure our belongings and personal safety. We take an umbrella with us when we suspect rain. We insure our homes even though the risk of fire, flood, or earthquake is statistically low. Farmers get insurance against crop failure. Investors seek protection against market slumps. Security can mean freedom from simple inconvenience to freedom from life threatening calamity.

When I worked for the EPA, we performed risk assessments to measure what exposure to a particular substance would mean to health and the environment. As a compliance investigator I later found that many corporations had an entirely different definition. Risk assessment might consist of calculating the odds of getting caught, or whether a penalty would be more than an inconvenience. Pollution might be a risk to the public, but if EPA's authority didn't have the muscle to be a risk to the business causing the pollution, there was no leverage for the company to clean up its act.

I had a tough lesson in risk assessment and the precautionary principle while leading an expedition in the Himalayas. At twenty thousand feet, when things go south you can't call 911 and wait for the ambulance. An expedition is a self-contained, two-way journey with a task in the middle. Resources, more than skill, often determine success or failure. Without diligent planning and oversight, surprises can threaten everyone. The more potential danger an expedition has, the more precautions must be taken in planning and execution to avoid unnecessary hardship and black swan events.

Kanchenjunga is the third highest mountain in the world. This enormous massif sits near the border where China, Nepal, and India meet. Our group was three weeks into an expedition to explore a remote valley between Kanchenjunga and the Chinese border. The precautionary principle compelled us to plan for the worst, hope for the best, and expect the unexpected. When a hundred miles from the nearest dirt road, food, water, and shelter are things you zealously protect. But there will always be unknowns.

That happened one frosty morning half an hour before sunrise when I heard someone unzip the entrance of my tent and felt a hand gently shake my foot. My head cook Galzin poked his head in and whispered, "Bara sahib (boss) there are two men planning to steal supplies. They tried to bribe porters to go with them. They want to take a tent, food, and bedding." I instructed him to fetch Lopsang, our Sirdar (foreman), and the three of us would handle it at breakfast.

We were weeks away from any help if we got stormed in or had a serious accident. We had planned to be self-sufficient and ready for most emergencies, but if we lost food or shelter, the entire team's security would be compromised. This was truly a black swan event that threatened everyone.

One of the men was a young entrepreneur of comfortable means, and the other was a high-ranking aeronautical engineer in his early forties. The two men were essentially trekking clients, and not veteran mountaineers. They apparently didn't understand the importance of limited resources or functioning as a team in remote regions.

When we all assembled at breakfast, I explained how important it was for the entire team to keep supplies secure. Then I reported that two of our team wanted to steal supplies, commandeer porters, and take off on their own. After my announcement, neither the team nor the two men wished to remain together. One team member said it was like a soldier stealing ammo from others before a fire fight. There was mention by some of the team to put the two in chains.

We were luckily able to obtain several porters in the next village. I encouraged the men to leave the expedition for everyone's safety. They could take the tent they had been using after making a deposit equivalent to its replacement value. I would return their deposit in Kathmandu after I received the tent back in good condition. They would have to pay the porters and find food and other provisions on their own. The rest of the team pushed on for a successful reconnoiter of the hidden Ghunza valley.

As I reflected on that experience, it appeared to me that the difference between an expedition to some remote region of Earth and a sustainable civilization may only be one of scale. Both have finite and renewable resources that must be protected. If food, water, and resources were compromised all other priorities are soon forgotten. Even with the best planning, human behavior remains the single most critical component of success. The unpredictability of human behavior is a pivotal factor in addressing the global crisis of climate change.

If we look at the relationship between security and the goods and services provided by the environment, two perspectives come to mind. On one hand, insecurity of natural resources and environmental conditions may contribute to conflict. On the other hand, natural resources, and the environment are negatively impacted by conflict.

According to David Jensen, Head of the UN Environmental and Peacebuilding Program, human behavior toward the environment is particularly important today. Nearly all international human conflicts or potential conflicts have a natural resource or environmental component. Environmental resources can serve to fund conflicts or fund the means to defend against those conflicts. Mineral extraction and deforestation are good examples. Oil has been at the center of conflicts around the world for over a century. Wars were fought to get oil and oil was sold to fund wars. Environmental management and policy can support peace. Poor environmental management can detract from maintaining the peace. It may also support or detract from successful reconstruction after a conflict. In either case, human behavior remains the lynch pin to success or failure. Today human behavior comes down to politics.

The economic value of minerals amplifies the risk of conflict. Illegal gold mining funded criminal groups in the Congo and Central America. In Sierra Leone, the economic value of conflict diamond mining continues to propagate human rights abuses. The primary motives for deforestation are agriculture and timber profits. Deforestation provides a vital economic benefit to local economies. At the same time, maintaining the forest has an international environmental value as a carbon sink, by capturing carbon and providing oxygen to the atmosphere.

The Amazon basin is often referred to as the lungs of the planet. It provides critical habitat for over half of all land-based plant and animal species. Many of these plants and some animal species have proven to be of major importance to medicine and the treatment of disease. The rain forest is the home of indigenous people who are threatened with extinction when economic motives remove their livelihood. The increased global appetite for meat and agricultural products increases the economic pressure to clear more land. Human behavior and the potential for black swan events lurk in the fine print of all environmental negotiations. The yin and yang of environmental security is as delicate as it is complex.

David Jensen reports that nearly half of all armed conflicts have either been financed by natural resources or to get financing for other conflicts. Most conflicts in the past two decades have had a direct link to changes in the environment or the mismanagement of environmental resources.

On January 2nd, 2016, a far-right extremist group led by Ammon Bundy occupied the Malheur National Wildlife Refuge in Oregon State. They wanted grazing rights on protected refuge land. A standoff took place between the Bundy group, and federal and local enforcement officers until February 11th, when most of the militants surrendered or withdrew. Several leaders were arrested. Robert LaVoy Finicum was shot and killed while reaching for a concealed handgun.

The economic value of natural resources may also conflict with cultural values. The Dakota Access Pipeline had a short-term economic value to Canada and the US but would seriously threaten First Nation cultural values and water supplies. After years debating in the courts and demonstrations between cultural and economic interests, in June of 2021, the Biden administration withdrew the pipeline permit honoring native sovereignty and treaty rights. In recent years, economic pressure resulted in proposals to mine in or near national monuments, parks, and wilderness areas. Mining proposals near the Grand Canyon would threaten one of the natural wonders of the world and further deplete water from the Colorado River.

According to UN Secretary General Antonio Guterres, "Many conflicts are triggered, exacerbated, or prolonged by competition over scarce resources; climate change will only make the situation worse. That is why protecting our environment is critical to the founding goals of the United Nations to prevent war and sustain peace." Maintaining the peace has the highest priority today. There are two man-made threats capable of destroying nearly all life. Nuclear war is harmless until the moment it ignites, but just as catastrophic is the existential progression of extinction. The only difference between the two is a quick death by holocaust, or a more protracted and agonizing extinction from environmental chaos and privation. Both provide mutually assured destruction (MAD). Neither has any rational justification.

An international forum first addressed this reality in 1972 at the UN Conference on the Human Environment. Soon after the Chernobyl nuclear disaster in 1986, Soviet Premier Gorbachev called for an end to Cold War thinking in order to take up the challenge of new environmental and planetary system risks. By 1987, the World Commission on Environment and Development (WCED), led by the Prime Minister of Norway Gro Brundtland, defined and popularized the concept of sustainable development.

From that time to the present, three new areas of security research began to develop. Environmental scarcity like soil erosion, water scarcity, overfishing, and deforestation were recognized as a growing source of grievances that could trigger conflict. A second area of research focused upon the distribution of natural resource abundance and how that relates to the equitable distribution of wealth and wellbeing. The third area of research centered on development. Resource rich nations may not be capable of economic development because of economic or political inadequacies. Corruption, weak governance, or a tendency toward short-term and debt driven development could increase social fragility and risks of civil conflict. Resource scarcity, equitable distribution and development are now at the forefront of environmental security research.

A sociopathic economic mindset had a strong influence on how wealthy nations viewed the resources of other nations. Fossil fuel exploitation is a prime example. Industrialized nations need an abundant supply of energy to drive the engines of their economy. In the twentieth century fossil fuels provided that energy. Wealthy nations and powerful multi-national corporations took advantage of weaker nations by offering technology, financing, and even bribes. This often resulted in economic imperialism by force and deception. Great Britain and France understood the value of plentiful, inexpensive petroleum to power their economies. In the first decade of the twentieth century, their eyes turned to the Arabian Peninsula.

Oil was discovered in the Middle East in 1908. Great Britain and France wanted to control those reserves. On the sixteenth of May 1916, Francois Georges-Picot of France, and Mark Sykes of the United Kingdom secretly agreed to spheres of influence. The Sykes-Picot agreement was an example where economic, political, and military motives established a pattern of deception and manipulation that have continued to this day. That hegemony was accomplished with the endorsement of the Russian Empire and Italy. At the beginning of WWI, a serious problem had to be overcome to accomplish the agreement's objective.

The Ottoman Empire (Turkey) had military control over most of the region. However, they had little control over indigenous tribes. Native tribes had controlled the Arabian Peninsula for thousands of years. Tribal territories were based on tradition, centuries old trade routes, and water rights. In covert support of the Sykes-Picot agreement, a lie was perpetrated to unite the tribes against Turkish forces. Allied western forces made promises to support the establishment of a pan-Arabian nation (Arabia) at war's end, in exchange for tribal support against Turkey. Unifying the tribes would not be an easy task.

Some tribes were openly hostile to each other. The western alliance needed someone with detailed knowledge of desert warfare and an intimate knowledge of tribal and religious customs. That task was assigned to warrior/scholar T. E. Lawrence (Lawrence of Arabia). At the beginning of the war tribes fought over water, then they fought for Lawrence and a Pan-Arabia. By the end of the war, they discovered the fight was over oil to power European economies. Lawrence brilliantly accomplished his task, though there is some question whether Lawrence was fully aware of the scope of the Sykes-Picot deception. Allied support for a Pan-Arabia was withdrawn at war's end. New nations were formed according to western alliances, not tribes, desert trade routes, or water rights. European powers made secret deals to decide who would lead these new nations. This reset tribe against tribe and a lasting mistrust of the Western infidel.

One of the greatest causes of conflict is privation. Over the past century fossil fuels have funded the accumulation of great wealth for the developed world. It has also been the cause of widespread poverty, war and chaos for others. The legacy of the Sykes-Picot Agreement was a lingering hatred between Arab peoples and western involvement in the region. A century of economic imperialism and Islamic resentment came to a head with the beginning of America's longest war on September 11th, 2001.

While western nations fought to maintain their oil economies, GHG emissions were destroying the economies of other nations. The 2010 Arab Spring uprising and the failing economies of Syria, Yemen, Libya, Somalia, North and South Sudan, the Democratic Republic of Congo, Chad, Pakistan, Nigeria, Iraq, Afghanistan, and North Korea have resulted in millions of deaths and tens of millions of refugees. Millions more clamored at incredible risk to escape scarcity, corruption, and conflict. Venezuela, Nicaragua, Mexico, Guatemala, and Honduras are countries in America's own backyard that are experiencing climate emergencies, corruption, and instability.

As these crises mount, powerful nations will continue to have food and resources simply because they can, but at what price and for how long? When a few come to your door, you may invite them in as guests. When millions seek shelter, you feel invaded. Refugees migrating to Europe are already seen as invaders by nations that greeted them with open arms only a few years ago. We can now see clearly how the connections between the environment, food, and water scarcity contribute to conflicts and mass exoduses around the world. The invasion by tens of millions of desperate climate refugees is just beginning.

Today the economies of all nations are linked. If one fails, the ripples spread throughout the globe. In 2010 and 2011, the world price of wheat and rice exploded. Extreme heat, fires, and floods in Russia, Pakistan, and Southeast Asia caused massive crop failures. This caused the global price of wheat, rice and grain crops to hit an all-time high. Global emergency relief reserves were quickly depleted. The February 24th, 2022, Russian invasion of the Ukraine threatened nearly 30% of global grain production. There was an immediate domino effect on the food security for nations already struggling to keep their economies afloat. For nations like the Democratic Republic of Congo, Yemen and the horn of Africa, a 30% increase in wheat prices meant the threat of famine for tens of millions.

Food shortages aren't the only consequences of climate change. The entire world is experiencing heat events, droughts, fires, storms, and epic floods every year that were previously once-in-a-century events or unprecedented. Nearly a billion people around the globe are living on the edge of starvation. Organizations that study security report that food supplies and economies will only experience increasing stress. Nations that attempt to protect their own security may threaten the security of others.

Ethiopia is nearing completion of the Grand Renaissance Dam on the Blue Nile. The dam is the largest in Africa. Who gets the hydropower and who gets the water to feed a drought-stricken region already locked in civil wars and starvation? Similar stress is taking place in Nepal as China and India vie for the hydroelectric potential flowing from the Himalayas. South Africa is facing the worst drought in a century. In 2015, water levels behind Cape Town dams declined 72%. Drought conditions have continued to worsen.

Twice in twelve years (2010 and 2022) monsoon storms put a third of Pakistan underwater. In the same period, China and India experienced record flooding and droughts. Record Yangtze River floods occurred in China in 1998, 2010, 2020, and 2022. Until recently, floods of this scope were thought to be once in a millennia event.

The Colorado River supplies critical power and water to five western states. Lake Mead is at the lowest level since Hoover Dam was built. Hydrology studies estimate that water levels will cease to drive the dynamos around 2030. What will happen to Phoenix, Las Vegas, the Imperial Valley, and Los Angeles then?

The African continent has the highest birth rate in the world. Africa is also experiencing some of the most extreme heat and drought stress. Increased drought conditions are the trend in Africa and America, but flooding can also occur when a decade of precipitation is delivered in a few hours as it was in Death Valley in August of 2022. Climate extremes continue to set new records. The trend is for more uncertain water availability and increased crop failures.

Scientists tell us that the worst mega-drought seen in fifteen hundred years has already begun in the American Southwest. Where will they get water when rivers become vernal streams and wells go dry? Where is America's resilience if beef and grain production is cut in half? America's rainy-day fund of natural capital is being withdrawn on a massive scale. The Dust Bowl was only a small sampling. The effect on agriculture and the economy has only begun to hit the average citizen's pocketbook. America now faces a new Grapes of Wrath as the Southwest dries up and begins to blow away.

Health security is another risk. Europe is gradually closing its borders as African and Middle East refugees huddle in densely populated camps, unable to move ahead and afraid to go back. These people are forced to live in crowded and squalid conditions. Those compulsory conditions present a breeding ground for disease. In the past decade, a dozen diseases nearly broke through efforts by the World Health Organization (WHO) and the Centers for Disease Control and Prevention (CDC) to stop them.

In terms of public health care, the US is approaching a third world nation status. The GOP's argument against single payer or an expanded Medicare coverage was that their health plan gave citizens freedom of choice. That was a ruse. There was no GOP health plan. Month after month, President Trump said it was on the verge of being released, but it never was, because it never existed. 'Freedom of choice' meant something entirely different to the average wage earner. If you had the money, you could choose the best health care. Those without the funds were free to choose limited care or no health care at all. Pay more and get less became the real health care pandemic in America.

At the time of this writing, there have been over seven million COVID deaths globally. The US and her territories have had more than a million deaths. Here is another way to think of it. Imagine that every man, woman and child in San Francisco just disappeared. Globally one in seven COVID deaths happened in the richest nation on Earth. One would think people would take that seriously, but our species doesn't think that way. We think in terms of 'me, right here and right now'.

A veteran Marine friend once related a story that illustrates this human trait. One day his platoon was being briefed about a very dangerous mission. The Officer in Charge said they could expect up to fifty percent casualties. Everyone looked at the guy standing next to them and thought the same thing. "That poor bastard." No one thought it would be them. When it comes to COVID or Climate Change, regardless of the objective statistics, no one thinks they will get sick.

The inability to see the relationship between climate change, economic instability, and the spread of hunger and disease is another crisis for global governance. Wealthy nations are not exempt from these interdependent insecurities. Hubris and ignorance have consequences. Even the richest nation in the world is far from immune to the shocks of disease, fires, storms, floods, droughts, or water and food shortages. COVID showed how vulnerable the world is. Tens of thousands of small businesses simply disappeared. America had neither the resilience nor the adaptive capacity to withstand even a relatively low mortality pandemic.

Agriculture was especially vulnerable. Fruit and produce growers require seasonal labor to live and work in substandard and dangerous conditions. Pork and chicken producers were compelled to consider euthanizing an entire year's stock. By the spring of 2020, the COVID pandemic caused the shuttering of several of the largest pork and chicken processing plants in America. Without labor for planting, harvesting, and processing many food producers found it necessary to let crops rot in the field or throw spoiled food away. Full recovery from COVID will take years and that future will never again look like the past.

When the world saw America's lack of resilience, its place in the global hierarchy declined quickly. Europe, China, and many Asian countries began transitioning their economies to build their own resilience capacity. America would be left behind with a weakened democracy, poor health care, a decaying infrastructure, and a century old power grid. America's security and global leadership faltered as competing nations stepped up to the challenge.

After eighty years of peace, another war in Europe added to the litany of security threats. American infrastructure was crumbling in place. We were no longer the world leader in manufacturing. Our old adversary Russia believed NATO and America were crippled by bickering and COVID. Putin felt it was a good time to invade Ukraine. The list of external and internal threats to America was growing. Wages were falling farther and farther behind inflation. Industry sought lower wages overseas and was now dependent upon temperamental supply lines to keep their profits growing. American was still dependent on the internal combustion engine to move things around.

For half a century America was the world leader in rail travel. In the 21st century, America's railroad system was left jogging in place while the world sprinted past. Rail transportation can move a ton of goods nearly 500 miles on a gallon of fuel. Highway transportation can move a ton of goods less than 100 miles on that same gallon of diesel. Even greater efficiency would be gained if rail transportation were electrified as they are in China, Europe, and Japan. In 2021, China had over 25,500 miles of high-speed track linking every major city in the country. That number is budgeted to double in the next decade. For more than fifteen years, China has had a maglev train that travels at over 400 kilometers per hour (~250 mph). America has less than fifty miles of experimental high-speed rail, and none in active service.

Marine shipping is a further example of America's declining economic security. Ninety percent of international trade moves over the ocean. Most US trade moves in vessels owned by other nations. According to the Department of Transportation and the Maritime Security Program (Dec. 2017) the entire US civilian merchant fleet consisted of only 393 vessels, or twenty-seventh place in the world. Russia was in eleventh place with 1,143 vessels, and China was second with 4,052 merchant ships.

Will America be able to survive a global military crisis with a foundering infrastructure, fossil fuel energy dependence, diminished manufacturing capacity, and an antiquated transportation system? What if Putin's illegal invasion of Ukraine expanded into a war with NATO nations? America would be called on to honor its NATO commitment. World War III would become a serious possibility.

Climate change has already had a dramatic effect on America. Depleted soils, water availability, and storms continue to reduce available domestic food production capacity. Florida has lost 90% of citrus production due to climate change and disease infestation. Beef production in the US depends on annual feed stock and water availability. According to the USDA, America's appetite for meat exceeds domestic production. That means 10% to 20% more meat must be imported to keep up with demand. Approximately 30% of fresh vegetables are imported. More than 75% of the fruit reaching American tables is imported (US DA, NY Times 2022). The linkages between nations for food, building materials and manufacturing make it nearly impossible for any single nation to sustain itself in an abrupt global catastrophe.

After Russia's invasion of Ukraine, people are thinking about the potential for another world war. Vladimir Putin has threatened the possible use of nuclear weapons. The Alliance for Science organization calculated that a limited world war with only one hundred nuclear detonations would bring about a global famine. The damage to the environment would be so severe that contaminated water, soil, and habitat would compromise the ability to feed the survivors. Russia has 4,477 nuclear warheads. The US has 5,500. None of the developed nations, including the US, could sustain a conventional 21st century world war. Billions would be trampled under the foot of military technology and armies taking priority possession of planetary resources. The residue of planetary resources would be insufficient to restart a thriving economy for generations, if ever.

Once the United States was the home of creativity and free enterprise. Adaptation and innovation were hallmarks of the American success story. Today innovation is dampened by poor governance and a lack of support for education, science, and innovative technologies. The US is failing to adapt while global security crumbles. Economic reporter Edward Alden puts it this way: "…while the US government was busy building the rules that unleashed the hyper-competitive global economy in which we live today, it did far too little to help Americans succeed and prosper in that economy."

Chapter 16
Greed, Sex, and Equity

I spent my life in science, exploring, teaching, and investigating violations of environmental laws. In recent decades, I'd seen America turn away from science and the rule of law. When I took up the task of researching global warming and sustainable development, I quickly learned that addressing them was impossible without understanding the economic and political factors that hinder sustainable progress and environmental security. It was also necessary to understand how so many accepted leaders and officials that blatantly and repeatedly lied to them.

In 2019, The World Economic Forum identified climate change as '…a major risk to good development outcomes'. It has become clear that there is the deliberate effort to support fossil fuel energy dependence, despite categorically overwhelming evidence that it is destroying national and global security. Donald Trump's irrational behavior in the face of this crisis has irrevocably and directly compromised the health, security, and prosperity of the US. The ripple effect of his behavior has directly impacted global security.

Fossil fuels have dominated the entire global economy for over a century. Fluctuations in fossil fuel prices are now the major contributor to boom/bust cycles, and economic destabilization. Yet nations continue to fight over oil. For the first two decades of the 21st century, Russia has had the EU and NATO countries over a fossil fuel barrel. The EU is now pulling out all stops to transition to sustainable sources and regain their energy security. Energy independence will give the EU and NATO the leverage needed to discourage further Russian aggression. In the long term, transitioning to sustainable energy will stabilize energy availability and reduce the cost of energy as well. The cost per kilowatt of sustainable energy has gone down 40% for wind and over 80% for solar in just the last fifteen years. Most of these reduced costs resulted from market growth, increased innovation and technology development.

There will be libraries filled with books written about the past few decades. Evidence of corruption was not hidden. It was broadcast on home TV screens, on the radio, the web, and in the press. That corruption of cultural and legal norms made governance ineffective. America's confidence in its own democracy was tarnished. In an era of multiple existential crises, intimately tied to a collapsing global habitat, the Trump administration acted to amplify that threat. Lady Justice took off her blindfold and was holding out her hand to the highest bidder. In a time of imminent crises, political gaming fomented chaos. What I was trained to view as crimes, were committed unabashedly and in plain sight. It was evidently more important politically to impress the bosses of industry than ensure the security and wellbeing of the electorate.

Despite Trump's denial of the science, climate alarm is rational. The US DoD says climate change and environmental disruption are threat magnifiers affecting every aspect of national defense, resilience, adaptive technology, and mission readiness. The interaction of population growth and a deteriorating environment objectively magnifies the tangible Malthusian troika of war, pestilence, and famine.

In 2019, at the UN Conference of the Parties (COP-25) youth activist and climate change flag bearer Greta Thunberg took the podium. With fist-clenched outrage, she declared that good intentions and endless conferences were not enough…action was everything. Her dismay did not come out of thin air. At her side sat the representatives of a growing alliance of millions of young people, and many tens of thousands of scientists, activists, and economists from around the world. Fear, even panic was justified, but that energy would have to be focused into tangible action. Inaction will never be forgiven. Every year news from the field is increasingly gloomy. Decades of modeling and predictions are exceeded with disturbing regularity. Extreme environmental events are already causing crop failures, arable land loss, unprecedented fires, floods, droughts, storms, and the crippling cost of recovery and growing insecurity. The demand for water, food, and energy is already testing the limits of human resilience. The human population, consumption, and pollution continued to increase.

Economists and health experts said that the impact of a single 'black swan' event could halt the entire global economy in its tracks. They said a pandemic was inevitable. Science warned that all the components of a sustainable civilization were inseparably linked together and limited by a changing environment. A major event in one component would impact all the others. The effect of a sudden change could trigger a domino effect on every sector of civilization. Experts gathered from around the world to sound the alarm. The time for action was so immediate it was difficult for many to believe. The actual collapse of civilization was not only possible, but it was increasingly probable without drastic and immediate universal action. Their reports were direr every year, then every month, then every week. The Trump administration and his control of the Republican Party ignored the evidence.

The threat of climate change on the human habitat moved security to the forefront of international concern. Climate extremes were a major factor that contributed to the chaos in Central America, the Middle East, and African nations. History clearly illustrated that environmental destabilization would spill over to other regions. In Sub Saharan Africa, climate change was making it more difficult for the land to support an exploding population. In repeated risk assessments, the DoD told the Congress and Trump, that after nuclear war, climate change was potentially the greatest threat to global security.

In late 2013, President Barack Obama signed Executive Order 13653, *"Preparing the United States for the Impacts of Climate Change."* Firmly backed by science, it instructed all federal agencies to identify global warming's most likely impacts on America's future and to take what actions were necessary to improve national preparedness. In March 2017, Donald Trump rescinded that order, and instructed agencies to abolish any rules or regulations adopted under the order and abandon their efforts to mitigate climate change. This action alone moved the US at least four years behind the curve to head off global climate catastrophe. Trump may have signed the death warrant for billions as the window to environmental security closed.

Everything the US National Academy of Science, thirteen US federal agencies and the DoD warned about was coming true. In nearly all cases, change was faster and worse than predictions. The close interaction between an expanding population, a collapsing environment, and fractured security became more urgent every day. Climate change means no one is safe. The world has played whack-a-mole by addressing one crisis after another with half measures. Broad holistic programs that included all three components of sustainable development were abandoned under the Trump administration.

Recently Donald J. Trump was arrested and indited for dozens of criminal acts that violated state and federal laws. He was accused of acts that violated the US Constitution and threatened American democracy. He could potentially spend the rest of his life in prison if convicted. However, his greatest crime was committed against humanity and future generations who will inherit an uninhabitable planet and an economy unable to salvage a secure future.

Hubris is a fatal flaw when a nation's economy relies on imports and ignores the loss of independent national resilience. Two world wars made Great Britain abundantly aware of this vulnerability. Germany's submarine blockade made starvation more of a threat to England than a Nazi invasion. It is increasingly clear that global food production cannot keep up with population growth as the climate becomes less stable. Something must be done to stop and then reverse population growth.

The world continues to face the crisis of too little food, for too many people, on too little land. Intellectually most understand the relationship between natural resources and the population. Politically, and in small highly vocal sectors of society, that reality hasn't gained much traction. New techniques in agricultural production will not help for long if the population is not controlled.

To most people it is abundantly clear that it is impossible to address overpopulation without addressing equity and reproductive rights for women. Humanity must control its numbers or sentence civilization to the endless ravages of war, pestilence, and famine. That reality sets the priority. Regardless of mores, universal education, female equality, and family planning remain the three most critically important factors in addressing the population crisis.

Throughout the animal kingdom, it is the female, not the male that has control over reproduction. It is the female's biology that determines when and if fertilization will occur. Her judgement and biology will determine if the fetus develops and grows to maturity. Only the female has the most detailed and intimate knowledge of her unique circumstances, not males or society. The party of Trump (GOP) and a Trumpian appointed Supreme Court have moved to remove female autonomy over conception, gestation, birth, and nurturing.

Sex for *Homo sapiens* is at least as important for social bonding as it is for reproduction. When society removes female reproductive autonomy, it breaks that social bond. In affect it legalizes rape by removing the female's control over conception. Western civilization has constructed a culture of female sexual exploitation and reproductive disfunction that aggressively contributes to the suppression of women. Male sexuality is as likely to be an expression of power as it is for sexual gratification or bonding with a mate. Testosterone tends to amplify character traits. Peaceful cultures tend to channel male aggression toward more nurturing roles. The US expression of religious dogma has tended to promote male dominance and keep both sexes profoundly ignorant of biology and sexual psychology. The US economy has commodified female sexuality by separating it from her human rights. Female sexuality sells everything from cars to household appliances. At the same time, it has reduced female authority over her body and legitimate role in society.

Women with reproductive autonomy tend to have fewer children. The Trumpian GOP and unregulated free-market capitalists argue that more people translate to more consumers and greater cash flow. They argue that a constantly growing population is needed to build a larger workforce to support social programs for the very young and elderly members of society. A growing population creates more consumers. In that respect, the Trumpian GOP view of female reproduction serves the same antiquated function it did for the extinct civilizations of Sumer, Rome, and Egypt where women served as free domestic labor and a continued source of manpower to protect the treasures of the rich and powerful.

Not all cultures believe in male dominance over the female. Native Americans have been around for more than twenty thousand years. Traditional Navajo culture is more egalitarian. Women have tended to play a major role in tribal governance and control most of the material wealth in the home. When I taught science on the Navajo reservation, I noticed how their matriarchal tradition had a way of holding on despite Anglo influences. Off to the side of a tribal council meeting I noticed an imposing woman weighted down with traditional silver and turquois. When a resolution was brought to the table and presented to the Chairman, his eyes would pause on that same woman. There would be an almost imperceptible nod of yea or nay before business continued.

A wise Navajo medicine man once told to me that the Anglo-American idea of economic progress went something like this. "If there are three people in a canoe with three peanut butter and jelly sandwiches, that is good. But ten people with twenty peanut butter and jelly sandwiches is better." At first, I didn't understand. He chuckled a little and explained that the problem was that the canoe hadn't gotten any bigger. Neither has Earth.

In 1999, Paul Hawken and Hunter Lovens wrote the book *Natural Capitalism*. "Ultimately, however, the chain leads back to biological systems, the sphere of life from which all prosperity is derived." The population is directly linked to the exponential growth of consumption and limited by that consumption. Unregulated free-market theory fails to consider that the environment provides literally 100% of the supply side for the economy.

India is the most populous country in the world with one point four billion people. It has the fourth largest national footprint, but their per-capita footprint is small. The per-capita American footprint is nine times greater. The national footprints of India, China, Indonesia, the EU, and the US collectively consume nearly twice the carrying capacity of the planet, though their per-capita footprints are dramatically different. These nations must take resources from underdeveloped nations just to survive.

Here is a way to compare how industrialized nations exceed their just share of planetary resources. China has an ecological footprint of 3.2 hectares per-person but only has a national biocapacity of 0.92 hectares to support each person. That leaves a deficit of 2.28 hectares per-person that must come from somewhere else. The US has an ecological footprint of 8.04 hectares required to sustain one citizen, and a national biocapacity of only 3.45 hectares. Neither country has the biocapacity to sustain itself. Continued growth and prosperity can only be accomplished by acquiring someone else's resources. This may temporarily raise the economy of rich and powerful nations, but it reduces the capacity for developing nations to raise their own prosperity. This brings us to the questions of environmental and economic justice.

Chapter 17
Equity

Throughout history, the wealthy have often viewed the poor as an impediment. Early in 19th century Europe, it was believed that the poor were that way because they were lazy. The more industrious would work their way out of poverty. Punishing the poor would motivate them to work harder to improve themselves. This rationale led to the establishment of European 'poor laws' in the 1830s as a means of instilling discipline.

In the US, conservative members of Congress continue to argue that welfare only perpetuates poverty and a lack of desire to better oneself. They apply this principle to giving aid to other nations and to immigrants, minorities, and women as well. It is true that a small minority will milk the system, but the numbers do not support the assumption for most disadvantaged people.

The Trump administration placed more obstacles for the underprivileged to overcome by defunding nutrition assistance, trimming the food stamp program, and attacking the affordable health care act, day care, family planning, reproductive rights for women, and a quality public education for everyone. At some point, those barriers dampen even the strongest motivation to succeed and ignores the promise of an untapped intellectual and more flexible labor resource. The Trumpian Supreme Court continues subjugating women.

When I was a young man, my father demonstrated how the will to try could be stifled. One day in my early teens, I wanted to buy a book. It cost a bit more than I had, so I asked my father if he would loan me a dollar. He held out a dollar. I reached for it, but he quickly closed his fist and pulled his hand away. He extended his hand again. Again, I reached for the dollar. Again, he jerked his hand away. Once more he extended his open hand, seeming to offer the dollar. This time I didn't try. My father told me that the world will often place obstacles in my way but it's important to keep trying. In the end, my father gave me the dollar. This was a particularly important lesson for me because the book was *The Sand County Almanac*, by Aldo Leopold. That book helped me decide to become an environmental scientist. That single dollar of assistance changed the rest of my life. It was important to keep trying, but a little assistance made my efforts reap an even greater reward.

Most people want a better life. Most will try, but if the opportunity is consistently withdrawn, eventually they stop trying. Social systems are intended to encourage pathways to progress and wellbeing, not inhibit them. Civilization has institutionally oppressed half of the human population's work force and over half of its intellectual capacity by holding women back. We do the same when we withhold assistance for the lower economic portions of our global civilization. There lies an enormous untapped reservoir of energy and talent.

When a tiny fraction of the wealthy fail to provide their fair share to the overall wellbeing of society, two outcomes become more likely. The have-nots may give up trying, or eventually a revolution takes place. The history of poverty and revolution clearly illustrates that the probability of one or the other increases as the gap between the haves and the have-nots widens.

The economist Thomas Piketty wrote in his meticulously documented opus *Capital*, that an unregulated free-market capitalist economy naturally produces a gap between the rich and poor. That gap will exponentially widen over time. Either conservative politicians fail to realize how wide that gap has become, or they are doing it intentionally. I tend to suspect the latter.

Business Insider magazine calculated that Jeff Bezos has two million dollars for every dollar the average citizen has in their bank account. He makes as much in twenty seconds as the average American makes in a year. By 2023, that twenty seconds was cut in half. In 2017, an Oxfam study reported that eight people, six of whom were Americans, have as much combined wealth as half of humanity. That's a big gap. At some point, the bottom 99% must be wondering, does the 1% have the right to decide who eats?

We are missing a wellspring of intellect and creativity hidden in women and the downtrodden classes of society. How many great minds, artists, and leaders rose out of the lower economic layers of society? Abraham Lincoln, Henry Ford, Oprah Winfrey, Starbucks' Howard Schultz, Leonardo da Vinci, Enrico Fermi, Picasso, Michelle Obama, and author J. K. Rowling were not born to great privilege. How many more might appear if given an equal opportunity?

That question was answered for me when I was an EPA liaison to the United Nations and World Bank. We assisted a few dozen nations in developing their environmental protection and enforcement programs. One day I was escorted to a cultural heritage center a few miles from Jakarta on the island of Java. There were reproductions of various native villages from the many cultures and islands of Indonesia.

My guide and I came upon a small, elevated platform with woven reed walls and a thatched roof. On the platform, a man was sitting cross-legged like a Buddha. He was naked except for a loincloth. His hair was matted and caked with mud. A six-inch chicken thigh bone pierced his nasal septum. He was making a blowgun. One end of a four-foot-long piece of bamboo was held between his toes while he carved the other end.

I asked my guide if the government provided assistance to indigenous peoples.

To my sudden amazement the native fellow said, 'Bloody damn little', in perfect British dialect.

"You speak very good English," I replied.

"Four years at Cambridge helped a little," he said.

"What are you doing here?" I asked.

"I'm trying to get the government to stop a mining operation that is killing my people."

He explained that he was a tribal chieftain sent by his people to complain about highly toxic mining waste that the American Freeport McMoRan mining company was dumping into his river. The river was his people's only source of fish, transportation, and water. Toxins from the mining waste were killing their animals. Fishing had completely stopped. Dozens of his tribal members had died and women were having miscarriages and malformed babies.

We spent the entire rest of the day discussing pollution, environmental science, economics, education, and politics. That naked 'savage' with a bone in his nose, was one of the most articulate and brilliant men I have ever met. The raw material of his genius was only brought to the surface by the opportunity for an excellent education. Of the planet's eight billion people, only a tiny fraction is granted the opportunity for higher education. How many more potential geniuses are we missing by not investing in a universal, high-quality education?

For years, I wondered if my friend was successful in fighting a system that valued profit over human lives. My question was answered in 2018 when I read that Freeport-McMoRan revenues totaled $18.6 billion, much of that coming from copper and gold operations in Indonesia. Had the ball mills of mining crushed another indigenous people into the mists of history?

Arcane economic practices further divide the 1% from the 99%. Those practices threaten the intellectual capacity of humanity and lower the ability of civilization to adapt to a rapidly changing environment. Trickle-down prosperity doesn't work and never did. It turned out to be just another form of *noblesse oblige* where the rich maintain control of their beneficence. The 1% keep it while some future Einstein sits like a Buddha, with a bone in his nose, carving a blowgun.

America is currently the wealthiest nation on Earth, but we've already established that isn't an accurate measure of overall prosperity. US production per-capita is among the highest in the world. To achieve that record, Americans currently work longer hours and have more stress related health issues than they did fifty years ago, and yet the actual buying power of wages has gone down significantly.

Despite the hype, America is moving backward. Tourists from China, Japan, and the EU are shocked when they land at a US airport. They see a nation in decay, with crumbling infrastructure, collapsing social safety nets, and a third-world education and health care system that clearly favors the wealthy.

If foreign visitors follow the news, they might notice how corporate persons suspected of crimes have the option to settle for a penalty without declaring fault or going to trial. Corporate persons may have the option to declare bankruptcy or financial reorganization to escape overwhelming debt. It is not the same for carbon-based persons. For example, bankruptcy is not an option for student debt relief. According to Forbes magazine, the cumulative US student loan debt in 2020 amounted to $1.6 trillion, essentially commodifying knowledge and compromising future competitiveness. Clearly the biological person is at a legal disadvantage compared to the corporate person. How did this happen and what is it doing to democratic principles?

Credit buying has maintained the illusion of prosperity, but real equity has declined. A declining percentage of Americans will ever own their home or car free and clear. The national economy is living on credit and the illusion of prosperity. Real wealth in America rests in the people and the natural capital (environment) that provides the fundamental wealth of all nations. In the past, it was possible to exceed the limits of local environments by importing resources, material goods, and food from other parts of the world.

Shortages could be overcome when wealthy nations bought or intimidated their way out of scarcity. Those options are rapidly disappearing as we pollute and consume more of our global habitat. It's worth repeating that economies cannot consume their way out of scarcity.

Conflict often comes when some people or nations believe they have more rights or privileges than others. The practice of wealthy nations persuading poorer nations to give up their natural resources is a form of economic imperialism with a long history of alienation and conflict. At some point, developing nations will vigorously insist on taking control of their own resources and mapping their own future. Russia discovered this when they invaded Ukraine in February 2022. Ukrainians no longer wanted their resources drained by the Russian bear. Britain learned this on August 15, 1947, when India successfully demanded independence.

American exceptionalism postulates that the American form of government is uniquely virtuous and worthy of universal admiration. Exceptional is as exceptional does, to paraphrase Forest Gump. The US lauded itself as the 'shining city on the hill', 'the leader of the free world', and the 'indispensable nation', and then Donald Trump acted like a self-centered, spoiled brat on the world stage by shouldering his way to the front for a photo-op. It isn't America's form of government but America's success at governance that the world watches and learns from. As our global civilization struggles for its very survival, exceptionalism becomes an anathema to international cooperation. When a leader or powerful nation believes they have the right to function outside the norms or interests of others, cooperation begins to falter.

The Second World War brought America to the pinnacle of world power and international influence. More recently, many in Washington have assumed that leadership in a new world order is America's manifest destiny. Some of the wealthiest believe they have an inalienable right to dominate society, exerting their influence in the economy and politics. The disproportionate influence of the billionaire class has become a threat to the fundamental principles of egalitarian democracy. This kind of exceptionalism is evolving in two ways. The first is a political move toward vindictive authoritarianism (fascism), and the second is the establishment of a hereditary aristocracy where wealth and political dominance becomes intergenerational within a few elite families and fortunes.

The attack on the World Trade Center on September 11, 2001, illustrated how American exceptionalism can lead to conflict. As early as 1998, Osama Bin Laden repeatedly wrote to the American people, pleading for America to cease using its power to take the resources of poor nations. He said America plundered his country and further impoverished his people while obscenely enriching a tiny fraction of Americans. He begged America to wake up and have sympathy for the strife caused by Anglo-American economic imperialism.

I do not have any sympathy for his subsequent actions, but we cannot dismiss the fact that those who flew into the World Trade towers believed in his arguments strongly enough to die for them. Exceptionalism, like class or race is a set of prejudices without rational or scientific basis. Throughout history a belief in an exceptional class has contributed to bigotry and civil strife. The segregation of humanity by wealth, race, and gender is a cultural phenomenon. There is no scientific justification that one race, group of humanity, or gender is exceptionally blessed or cursed.

Children are born without prejudice. As children grow to become adults, cultural myths of race and class evolve to distort the way we view others. By the time we reach adulthood, many are totally unaware of the biases that culture has imprinted on their lives. Gender bias is one of the oldest forms of male exceptionalism. The fact remains we are all one family of the genus *homo* and species *sapiens* sharing a single habitat we call Earth.

Overpopulation occurs when the population exceeds the resources necessary to sustain it. For several hundred thousand years, the human population struggled just to survive. For most of human history, there were more resources than people needed. The UN determined decades ago that the best indicator of a stable nation is the status of women. This becomes more evident when the population is no longer sustainable. Today there are too few resources for too many people, but women are still without full reproductive rights. Overpopulation cannot be separated from women's rights. Fundamental to justice and the equality of the sexes is female reproductive autonomy. In the past, civilizations depended upon women to increase the population of craftsmen, laborers, artisans, and farmers to sustain society and to replace soldiers who died defending it. Gender inequality had more to do with maintaining the economy and male authority than male merit. Real power today has more to do with intellect and skill than testosterone.

The challenges we face today don't justify an archaic philosophy that half the population remain unfulfilled brood mares. We cannot address climate change or sustainable development successfully without the full participation of women in every facet of civilization. Women and children must be educated and given every opportunity to achieve their full potential. There are no exceptional people. We are all made from the same genetic fabric with interchangeable parts. No one should get a free ride on the shoulders of those who do the work. There are too many of us in the same canoe. In this time of existential crises, either we all bail, or we all sink together.

The global population has quadrupled since the beginning of World War II and will probably add another two billion more in the next thirty or forty years. Education, family planning, and female reproductive autonomy provide the most reliable means of controlling population growth. Sex is an extremely powerful motivating force. Most societies encourage self-control, but that has never proven to be an effective solution to birth control. Pregnancy can occur at any time and place, with or without female consent. Complex hormones and powerful human emotions may cause the moment to take priority over the consequences.

In the early 1960s, effective birth control caused a major shift in population demographics. The age distribution in developed countries shifted from a broad-based pyramid shape to one that is more columnar. This was thought to present several problems. One of the arguments against family planning was the fallacy that without a growing population it would be impossible to support a growing economy and social programs. That argument is not supported by the evidence. The Scandinavian countries, the EU, Japan, New Zealand and Canada all have low birth rates, but strong social programs, and growing economies. A columnar population can set aside enough throughout working years to carry it through retirement. This can be accomplished by several means. One system uses an Individual Retirement Arrangement (IRA). This would allow savings and investments to accrue with interest. Federally insured stock and bond investments can also provide a high level of financial security at retirement.

An unregulated free-market economy is based upon risk and interest. The Anglo-American market system promotes the concept of buy now and pay a lot more later. This artificially keeps the economy humming until the bubble bursts. Adapting our economy to changing circumstances is hamstrung by this arcane ideology. Saving for that rainy day is contrary to unregulated free-market economics though pay-as-you-go and saving for a rainy day is far safer and less subject to dramatic fluctuations.

No system is perfect, but some systems have shown more resilience and adaptability than others. It isn't the kind of government but quality of governance that counts. American politicians in both parties associate democratic socialism to communism, though most social democracies have clearly demonstrated a higher standard of living and greater economic resilience than the US. Scandinavian countries pay higher federal taxes, but on a per-capita basis still manage to save more, have better health care, maintain better public schools with much higher achievement scores, have longer life spans, and score higher on the overall happiness index than the average American. In those countries, corporations pay their share and don't have huge tax loopholes like those found in the US.

China is still considered a developing country even though it is the second largest economy in the world. They have worked an economic miracle over the past three decades. At its present rate of growth, it could overtake the US by mid-century. However, China is making the same mistake Britain and America made when they developed their economic empires. To sustain its remarkable growth, China is building its economy by taking resources from others. This is evident by their ventures onto the continent of Africa and nations on the rim of the South China Sea. China has claimed the Spratly Islands and is moving aggressively for greater control of Hong Kong and Taiwan. China claims the Spratly Islands are for peaceful purposes, though surveillance clearly shows intensive militarization. The UN, European Union (EU), and the United States have rejected China's claims, with growing international concern that China's actions are a precursor to further expansion.

Massive urbanization is another component of population growth that alters the calculus of sustainable development. In 1900, fewer than 10% of the global population lived in cities. Today, that number has grown to roughly 50% globally, and about 85% in industrialized countries. By mid-century, more than three-quarters of the world population is projected to be urban. Rapid urbanization creates both challenges and opportunities.

A problem arises when migration to the city takes place faster than urban planning and development. Hundreds of new megacities (cities of greater than ten million) will appear in developing countries over the next decades. Urban planning will determine if cities operate efficiently, or in chaos with stark segregation between the privileged haves, and the slums and favelas of the have-nots. The rising cost of housing leaves lower income people stranded in increasingly desperate conditions. If leaders anticipate this, infrastructure can be designed for future energy and resource use, waste management, and more efficient transportation. Low-cost housing can be developed. Infrastructure can be designed for the long term, and not the typical pendulum swings of election cycle politics.

Stresses will increase for rural areas. Demand for greater food production will place an ever-increasing burden on the farmer and the land. Urban consumers resist paying higher food prices, while the farmer faces rising costs just to maintain production. American farmers and ranchers make up less than 2% of the population. They receive less than seventeen cents of every dollar consumers spend on food. In good years, farmers may do well, but it only takes one or two bad years to threaten foreclosure. The rising cost of production places an even greater strain on the small farm. Their margins grow tighter as the cost of equipment, fertilizer, seed, fuel, and pest control continue to rise. For many, climate change spells the end of generations living on the land. The family farm is disappearing as they are absorbed by huge agribusiness corporations.

The corporate economy is transforming food production, but not in a good way. Massive agribusiness practices use more chemicals, lower natural soil fertility, and increase erosion. In many cases, genetically modified crops compromise nutritional levels in favor of production efficiency. A pleasing appearance can belie food quality. The produce aisle displays vegetables that look like they came off an assembly line. Tomatoes and other vegetables appear attractively uniform in size and color on the outside, but may be pithy, flavorless and less nutritious on the inside when compared to heirloom varieties.

Four federal programs have food safety oversight: The Food Safety and Inspection Service (FSIS) of the Department of Agriculture (DoA), The Food and Drug Administration (FDA), and the Centers for Disease Control (CDC). Every year regulatory oversight has greater difficulty administering the complex task of food safety. The inspection of food commodities is already less than most people imagine. For example, when the FDA certifies a food product to be *wholesome*, most people think it means nutritious. For fruit and grain products, wholesome may only mean that it has less than a specified percent of rodent fecal matter, hair, or bugs.

Another example is the meat processing industry. It is not uncommon for the meat processing company to pay the inspector's salary. Using the honor system to protect the nation's food supply does not have the best track record. Working conditions may become lax, corrupt, and overcrowded. There may be unsanitary worker safety and food hygiene practices. This was vividly illustrated more than a century ago in Upton Sinclair's 1906 novel, *The Jungle*. Some of those problems remain to this day.

In 2020, there was a COVID outbreak at the Smithfield pork processing plant in Sioux Falls, South Dakota. Poor hygiene and close working conditions allowed the spread of the disease in a mostly Hispanic workforce. Twenty years earlier, I inspected the same processing plant for the EPA Office of Suspension and Debarment and reported similar hygiene, inspection, and labor concerns. Conditions in agricultural work often demand long hours, under the most difficult and sometimes hazardous conditions. Housing for itinerant workers seldom meet the same standards found with more permanent employment. Agricultural chemicals pose another health threat to farm workers.

Nearly a fifth of America's food supply is imported. Imported food products may have less oversight. According to the FDA; "…more than two hundred countries or territories and roughly 125,000 food facilities and farms supply approximately 32% of the fresh vegetables, 55% of the fresh fruit, and 95% of the seafood that Americans consume annually. The increasingly globalized and complex marketplace has placed new challenges on our food safety system."

So long as it is available at the grocery store, the urban consumer pays little attention to how it got there, or what it claims to be. The hubbub of urban life leaves little time to contemplate food supply chains and the natural capital that supports it. Urbanization insulates the public consciousness. People lose sight of the critical value of soil and water and their fundamental connection to the climate that controls them. Food, water, and soil power civilization. Civilization is making more people, not more land. The demand for more food and water increases the pressure on natural resources to satisfy that demand. But it is the climate that ultimately determines production. An unstable climate causes crops to fail. Crop failures raise food prices and diminishes food availability. When droughts fail to provide water to grow crops, people go hungry. When floods destroy a crop before it can be harvested, people go hungry. Storms, cyclones, hurricanes, and the extreme temperatures of climate change impact food infrastructure. When infrastructure fails to prevent contamination, provide safe storage, or get food to market, people go hungry.

Chapter 18
Stewardship

Benjamin Franklin said, "When the well runs dry, we learn the worth of water." Civilization is running out of fresh, clean water. We know that less than 3% of Earth's water is fresh water. Only 1.5% of that 3% is potable water. The demands on that 1.5% of 3% has increased five-fold since 1900. It will increase another five-fold by 2050 but supplies will fall short well before that, according to domestic and international monitoring agency reports.

Approximately three fourths of all freshwater use is agricultural. It is projected that the food demands of a growing population combined with changes in food preferences will require a 75% to 100% increase in global food production over the next fifty years. This will place unprecedented demands on soil and fresh water supplies.

Why don't we just desalinate the ocean?

Desalination is relatively easy to do but requires huge amounts of energy for massive scale purification and distribution. Even osmotic systems require enormous facilities and infrastructure to bring production up to scale. How do we pump millions of acre-feet of desalinated water to interior farmlands around the world without breaking the bank? The cost of fresh and potable water is projected to run into the trillions of dollars by 2030. There is also the problem of equity. Water distribution tends to favor the wealthy. How is the task of freshwater distribution incorporated into economies that are more concerned about the quarterly report than long-term public needs or equity?

Can't we just dig more wells?

No—we are using up groundwater faster than it is being recharged. Some of the largest aquifers in America are already in imminent danger of contamination or running dry. Pollution, radiation, and fracking add to the problem. Even if we had more arable land we are running out of water for irrigation. We already see civil strife, refugees, and major security threats around the world that can be traced back to inadequate supplies of water. It is not clear to the general public that climate change is reducing the ability of the land to support the people already here.

Can't we just clear more land?

Globally, arable land is shrinking in area and fertility. Clearing more land removes forest carbon sinks, releases more CO^2, and increases global warming. More arable land is lost each year to poor farming practices, mining, pavement, and development. Substituting natural soil replenishment and pest control with chemicals causes runoff problems that lead to potable water contamination, fishery depletion, deoxygenation, and marine dead zones.

Can't we just grow more food?

There are cleaner, greener, and more profitable ways to use the land. This new field of agricultural science is called *agroecology*. Agricultural sustainability requires responsible stewardship of the natural systems and resources farms rely on. Soil stewardship rests on four core principles: build and maintain healthy soil, manage water efficiently, minimize all forms of pollution, and promote biodiversity. Some annual crops may be altered to become perennials that don't have to be purchased and planted every year. Several of these methods have already proven to increase production over conventional methods as well as promote natural soil fertility renewal. More importantly to the farmer, many of these methods have proven to be more profitable than conventional commercial farming methods.

Most of the resistance to new farming methods fall into three categories. The first is the need to improve education so that farmers understand the current research. Secondly, many of the new methods will require new farm equipment. That is an expense to farmers that may already live on tight profit margins. Lastly, farm loans and crop insurance will be necessary to cover lean years over the period of transition.

There is powerful corporate resistance to these more organic measures and practices. Some of the largest chemical and agricultural supply companies feel wide use of these new methods would hurt their profit margins. Companies like Monsanto lobby politicians and seek to control the research and protect seed and chemical patents. The economic connections between these chemical companies and destructive corporate agribusiness practices are well-entrenched and unlikely to change quickly.

Amery Lovins of the Rocky Mountain Institute has documented that a nearly 50% global footprint reduction is possible through improved efficiency and conservation, but only if the population remains stable. Even if strict efficiency measures were put into place, a growing population would still consume more resources. Sustainability, therefore, requires both a reduction in consumption and the number of people consuming. It's easy to forget that nature demands we play by her rules.

A corporation exists to make a profit, not to benefit humanity, but we can't put all the blame on corporations. That would be like blaming a lion for eating meat. It's in their DNA, or in the corporation's case, in their charter. Controlling when and how much the lion eats is the issue.

A small but influential portion of civilization is becoming lawless. Without oversight the people are blind. Without regulations there are no standards of behavior. Without enforcement there is no means of correcting bad behavior. History tells us when that happens the power brokers of special interest will take over. That is abundantly clear today. Without virtuous politics both society and the environment will continue to suffer. Today we face a crisis that is already destroying life as we know it. Governments that fail to form resilient, sustainable, and adaptive policies to address climate change, and environmental disruption are guaranteeing civil strife, conflict between economic classes, loss of fundamental resources, and the eventual collapse of civilization. That is the very definition of futurecide.

"The collapse of civilization can be avoided by reducing consumption and expanding women's role in society, biologists report" (Futurity, 2013). If we understand the ecological axiom that populations grow until they reach the limits of natural resources, wouldn't full speed ahead to please the captains of industry be a sure way to sink civilization? Wouldn't it be logical to take measures to control population growth and the economy so that it remained within the sustainable carrying capacity of the planet? Isn't that the definition of stewardship?

Chapter 19
Women and Civilization

Men may claim to have conquered the frontier, but women make it a home. Research by dozens of international organizations have established that, "Gender equality is not only a fundamental human right but a necessary foundation for a peaceful, prosperous and sustainable world" (UN). Elevating the status of women is accomplished through education, full equality in all aspects of society, and autonomous control of their bodies and reproductive functions. So why is the US moving backward?

Blatant gender bias appears in all economic and social classes. In law, politics, and most religions, men have falsely interpreted their role as the dominant gender. As a result, women continue to experience a gender biased 'glass ceiling' throughout much of the world. In some cultures, women are simply chattel with few rights or authority of their own. In a few, the husband may beat his wife—the justification resting solely at the discretion of the husband. Despite the strong traditions and cultural perceptions of the female role in society, there is no rational or biological justification for limiting female participation. Compared to other developed nations, the US has been incredibly slow to recognize those rights and potentials.

In 1880, the age of sexual consent in most states was ten or twelve years of age, except in the state of Delaware where legal consent was seven. In the US, women were not allowed suffrage until 1920. Women could be fired for being pregnant until 1971. Women could not serve on juries in all fifty states until 1973. Women could not obtain a credit card in their own name until 1974. Women were not admitted to US military academies until 1976. Women were not admitted to Harvard until 1977. Spousal rape was not criminalized in all fifty states until 1993. Companies were allowed to charge women more than men for health insurance until 2010.

Our global civilization is now zealously engaged in a new epoch of human power and discovery. There is greater promise today than at any time in history. Simultaneously the stresses of unparalleled change places pressure on our social structure and cultural norms. Humanity is facing the unforgiving limits of irrational growth. If we are going to have a just, equitable, and sustainable civilization, male aggression must be held in check. A measured and nurturing hand is needed on the hair trigger of self-annihilation.

For more than three hundred thousand years, we were assured that the human male was the hunter and the female the gatherer. We are now learning that was not always the case. Roles were not only driven by culture but also by hormones. Muscle mass and aggression are both strongly influenced by testosterone. The question we must answer now is what function does testosterone play in today's world? There are no cave bears or saber-toothed tigers to ward off. Somewhere between mitigating male aggression and the encouragement of female assertiveness lies a rational, more sustainable future.

The aggressive, testosterone driven male of history and fiction is no longer necessary for the delicate connectivity of an overpopulated, resource-drained, and lethally armed planet. I can hear the fists pounding on the bar. "Them's fight'n words. Only a wuss would come up with malarky like that." But do I speak blasphemy?

Human history is filled with old men pounding tables and marking their territories with the blood of young men and women. How many wars might have been avoided if women played a larger role in arbitrating the peace? That doesn't mean women are incapable of fighting. The human female is fully capable of showing her teeth when needed. Lakshmibai, the Trung sisters, Queen Boudica, Fu Hao, and Tomoe Gozen not only won wars, they led the charge. It's the female lion that feeds the pride. It's the momma bear that protects the cubs.

If we are to live sustainably as one people sharing a single global habitat, the territorial male must make way for women to become fully franchised partners. If this does not happen, we are likely to continue suffering the same injustice and inequity that brought us to our present predicament. The frontier has been conquered. It must now be made a home.

Men tend to overlook how significant the female contribution to civilization has been. Ask any realtor if they sell a new home to the husband or to the wife. Most realtors will say the wife. Women are more likely to vote for better schools and health care. Women tend to want better support for children and the disadvantaged. Women tend to be more supportive of social safety nets, childcare, and care for the elderly.

Physical strength and aggression traditionally tended to be the predominant argument for male dominance, but that is an extremely narrow view of human biology. Biological efficiency, longevity, reproduction, resilience, endurance, and a host of other measurements tilt slightly in favor of the female. Melvin Konner, MD, PhD. said, "Women are not equal to men; they are superior in many ways, and in most ways that will count in the future." William Golding said, "I think women are foolish to pretend they are equal to men, they are far superior and always have been." Perhaps men are afraid of something they have known all along. Socrates may have recognized this when he said, "Once made equal to man, woman becomes his superior."

Failure to prosecute violent assault against women is a further indication of a gender bias and male insecurity prevalent in 21st century America. Women often refuse to report assault or rape. They know the defense will typically attack her, not the perpetrator. Rape may be argued as assault with a 'friendly weapon' that was motivated by the female. In some cultures, the female may be held responsible for rape, because she had failed to adequately conceal her irresistible allure from masculine impulses. The defense might claim the male simply interpreted 'no!' as teasing consent. This would be the coquettish 'go away closer', 'stop it some more' defense. Domestic violence statistics show that women are far more likely to be killed by men, than men killed by women.

Legal protections for women from male violence have been sketchy at best. In 1991, attorney Anita Hill testified that Supreme Court nominee Clarence Thomas had sexually harassed her when she was his legal adviser. Ms. Hill alleged that Judge Thomas had groped her at a dinner party. Psychology professor Christine Blasey Ford alleged that US Supreme Court nominee Bret Kavanaugh sexually assaulted her by locking her in a room and pinning her down on a bed. In both cases, Thomas and Kavanaugh's defense was not a legal argument but a public display of highly emotional anger. Those displays were not the cool objectivity one might expect when under consideration for the highest court in the land. In both cases, the female complainants were belittled and treated with demeaning sarcasm. Nevertheless, both men were confirmed and sit on the court today. The relevance of these two nominations carries over to the present.

Mammalian reproduction is determined by the female reproductive cycle and the environment. I am unaware of any other mammalian species where the male has any authority over pregnancy or giving birth. The human female's right to reproductive autonomy is uniquely relevant to an overcrowded, resource depleted, and insecure world of climate change and environmental collapse.

The sole right to reproductive autonomy was decided in 1973 by the Roe v. Wade decision. In subsequent years, that decision was upheld by the Supreme Court at least seventeen times. In their confirmation hearings, both Clarence Thomas and Bret Kavanaugh clearly stated in their legal opinions that Roe v. Wade was established law. They lied. Both Kavanaugh and Thomas voted to reverse Roe v. Wade. That decision compromised the rule of *stere decisis* (the doctrine that courts will adhere to long established precedent in making their decisions). The decision to reverse Roe v. Wade removed a constitutional human right that had held for two generations. Women grew up with confidence that they had legal dominion over their own body and all its reproductive functions. That US Supreme Court decision opened the possibility to retry any other legal precedent from voting rights to contraception.

Sarah Wald, adjunct lecturer in public policy and HKS co-chair of the Joint Degree Program in Law and Policy wrote, "The decision (reversal of Roe v. Wade) also has potentially devastating implications for other rights, many longstanding and currently part of the social, historical, and moral fabric of America." By removing *stere decisis,* the balance of power between the branches of government shifted to the Supreme Court. As with fascist conservatism in the House and Senate, the Supreme Court is now clearly an impediment to the justice and equality necessary for sustainable development.

One of the primary forces behind the anti-abortion movement came from evangelical religious interpretations. Their positions tend to rest upon literal interpretations of relatively modern biblical translations of scripture. However, many religious scholars question modern translations. It's common to see religious subjectivity in philosophical debates on morality and ethics. Religion is frequently used by the church as the foundation of morality, but they are not synonymous. Religious philosophy is a subset of moral philosophy and not the other way around. Moral and ethical arguments behind laws and the norms of society can certainly be valid without any religious predisposition and far more likely to be founded on evidence and fact than ideological faith. When conservative politicians use religion as justification for bigotry and a moral basis for ethics, they abandon the staunchly held tradition to separate the church and state. Once political morality is tied to religion the question arises, which religion?

The GOP claims America has always been a Christian nation. That is not and never has been the case. The founding fathers specifically wished to avoid defining America in specific religious terms. Thomas Jefferson, George Washington, James Madison, James Monroe and John Tyler often identified themselves as deists. (Deism is the belief in a supreme entity based upon reason rather than revelation or the teaching of any specific religion, and that that entity does not act to influence events.)

The objection to abortion should be founded on rational evidence, not religious doctrine or belief. There is a recent renewed zeal in America to pass anti-abortion laws that would charge women and their doctors criminally responsible for a capital crime (murder) if they terminated a pregnancy for any reason, including saving the mother's life, incest or rape. Some of these laws would include an ectopic pregnancy where the fetus develops outside the uterus. Who has committed murder then, when an untreated ectopic pregnancy is almost certain to kill both the mother and baby?

Abortion is an emotionally torturous decision that involves a wide spectrum of personal, social, and economic moral considerations. Most of the US population would object to anyone else deciding what they should do with their home, personal belongings, or their person. Poll after poll indicates at two thirds of the US population was in favor of keeping Roe vs Wade. More than 85% of the population favors the right to abortion when the woman's life is in danger. Most of the US population favors a woman's right to choose before the fetus is viable. The debate then shifts to the definition of viable.

A baby is considered premature if delivered before 37 weeks of pregnancy. With intensive and highly expensive artificial incubation measures, a fetus may survive after 23 to 25 weeks of pregnancy. The medical costs to care for a premature delivery are unaffordable for most single women and many married couples. The cost for the postnatal care of a twenty-eight-week-old preemie in America is approximately $125,000.00. According to the National Library of Medicine and Translational Pediatrics, the average cost for postnatal care of a twenty-five-week-old preemie was nearly $300,000.00. Average daily ICU costs average around $3,500.00 a day. A prolonged stay of more than a month in intensive care can run to a million dollars or more. Current trend toward draconian anti-abortion laws forces these costs on both the mother and insurance companies. The impact on the medical insurance industry has increased the cost of health insurance for all policy holders across the board.

The point is that removing a woman's right to choose has many economic, political, environmental, and social consequences. Beyond the mother and child's future, it places an enormous burden on society that must be paid later. Those burdens will result in far more than one life and death decision. It will place a continuing burden on society, health care, the economy, politics, and the law. After weighing the lingering burden on the mother and an unsupported child to society, the vast majority of the US population favors a woman's right to choose.

Women's reproductive rights are the key to controlling overpopulation. But let's take a moment and discuss overall women's health care in the US. The treatment of women in the US has a terrible track record. In 2017, more women died in childbirth than in thirteen other developed countries. Family planning clinics inform expectant mothers of this and other concerns when they come in for counseling. Despite biased reproductive laws, domestic violence, gender bias in the workplace and a second-rate health care system, women in the US still manage to outlive men.

The more biologically robust and efficient female is tougher in other ways. Male babies are less hardy than female babies. Infant boys are weaker and more susceptible to disease than female infants. According to Science Daily, baby boys are 60% more likely to die in their first year than baby girls. Boys are 60% more likely to be premature and to suffer from conditions arising from premature birth, such as respiratory distress syndrome (RDS). Male babies are also at a higher risk of birth injury and mortality due to their larger body and head size. Premature males are more vulnerable and have a greater chance of problems later in adulthood.

I've seen men and women in the toughest circumstances after decades tramping wilderness trails. A standard porter load for a Himalayan porter is sixty-six pounds (~thirty kilograms). Men and women porters carry the same loads. Few things are more humbling to the hardy male Himalayan expeditioner than to stagger into camp and be handed a hot cup of tea by a petite female porter. The trekker probably only carried a light daypack, while the female porter trekked the same distance, arrived hours earlier, set up camp and put the kettle on. The 'client' typically found their tent pitched, sleeping bag and pad laid out, and supper well underway.

Women often prove equal, and occasionally superior, in traditional male activities. We see stellar examples in flying, racing cars, politics, and corporate management. There is some evidence that women pilots are capable of withstanding slightly higher 'G' forces than men. This is a distinct advantage when flying fighter planes. Women have proven to make fine astronauts. They are certainly competitive on racetracks around the world. Women often succeed in open gender competitions in traditional male sports like shooting and archery. No one would claim Annie Oakley (Little Sure Shot) couldn't hold her own with any man, in fact, she was never beaten in open competition.

Today there are more women in American universities than men. Despite this, it is not uncommon to find women with advanced degrees performing the most menial jobs. Thousands of years of cultural misogyny have ingrained a false perception of feminine capability at the sacrifice of half of the labor force and at least half of humanity's intellectual capacity. The United States is going through another spasm against women's rights. Sexual abuse is in the headlines, but male domination and the glass ceiling remains. Whether it is the interpretation of the law, cultural gender bias, or the interpretation of religious doctrine, men are still bending the rules to their benefit just as they did before the women's suffrage battle raged in the streets.

Today women have a powerful tool at their fingertips. The Internet and social media allow women to quickly establish networks and chat rooms on more than the price of groceries and diapers. Women are on the march again. This time they are educated and outnumber men. Women are trying to make this 'brave new world' just, equitable, and more sustainable.

Women are trying to make the coming world a home worth living in, if men will only get out of their way and listen. The climate crisis demands that everyone contribute. Leaving women out of the calculus is like tying your best hand behind you. Prime Minister Margaret Thatcher said, "In politics, if you want anything said, ask a man. If you want anything done, ask a woman."

This has been a difficult chapter to write. I know many men will take offense. Tough! I don't know how to be subtle about the truth. Most women will think there is nothing new here. I'm just an old rooster who has spent enough time with incredible women to see what I was too blind to notice in my youth.

Male aggressiveness really is a danger to civilization when weapons of mass destruction have replaced the fist and club. Women are well suited for full participation in the development of a sustainable world. Men should seek to be equal partners or the meek might just inherit the earth, or what's left of it. The planet and civilization need female nurturing and the male dominated world should make sure they get it. There is little doubt that it is also in men's self-interest. There may be another reason men haven't thought about.

Advances in genetic engineering (CRISPR) could make the male role of internal fertilization a quaint memory. Improvements in genetic engineering will soon make the male's 'Y' chromosome superfluous. It would be wise for males to make nice, lock arms with their better halves, and face a precarious future together as full partners. Listen up guys. Women are suiting up. I'm just saying....

Mark Twain summed it up when he said, "What would men be without women? Scarce sir...mighty scarce."

Chapter 20
Discounting the Future

Time and distance influence both our awareness of things and the priority we give them. The greater the distance from here, the less priority we place on a thing. The farther something is in the future, the less importance we place on it. These two factors make it difficult to communicate the multiple crises facing a sustainable civilization.

Discounting the future explains why governments are generally reactive rather than proactive. Senator James Inhofe threw a snowball in the US Senate to prove climate change must be a hoax. All he proved was that it was snowing outside. His immediate perception of reality was discounting future change. This trait of human perception poses a monumental psychological hurdle if society is to ever understand the multiple, evolving crises climate change poses.

This problem was illustrated in 1951 when the film *The Day the Earth Stood Still* was released. In this example of art imitating reality, an alien ship landed in Washington DC. A Martian named Klaatu was sent by a federation of planets to warn humanity that it was becoming a threat to peaceful planets. Klaatu's message was only for the Earth's leaders. If humanity failed to control their new knowledge of atomic power, a police force of all-powerful robots (Gort) would destroy all life on Earth.

The scope of Klaatu's power was demonstrated with a harmless example. He brought every form of technology on Earth to a standstill. Top scientists understood the gravity of the situation, but world leaders found excuses not to participate. Instead of listening they shot the messenger. Alien technology was powerful enough to resurrect Klaatu. For world leaders, their political reality seemed more immediate. Even after his figurative crucifixion, Klaatu hoped humanity would see reason. When Klaatu waved a casual salute and flew away, the movie audience was left with an unanswered question. Would world leaders realize the threat before Gort returned and turned Earth into a 'burned out cinder'? Civilization's ultimate destruction by climate change and environmental disruption poses precisely the same problem of discounting the future.

In 1980, the total concentration of greenhouse gases was less than 350 parts per million. Projecting that trend into the future, indicated that much of the world would eventually become uninhabitably warm. In the late 1980s, various conferences began to establish goals to reduce greenhouse emissions. By the 1990s, it became clear that the rate of greenhouse gas emissions had to peak as soon as possible. Global warming would soon cause catastrophic disruption of planet wide systems and compromise the fundamental supply side of the economy. The science was clear, but the public still failed to understand what it meant. Governments around the world discounted that future in their priorities.

At COP16 in 2010, the IPCC decided that 3.6°F (1.5°C) was the maximum amount of warming allowable to prevent an existential threat to civilization. At COP21, in December 2015, 195 nations signed the Paris Accord to do their best to lower their emissions with the global goal to remain below 3.6°F (1.5°C). To accomplish that goal, net emissions would have to be reduced 30% by the early 2020s and continue to be reduced until reaching net zero before the middle of the century. The sooner net zero emissions were reached the better.

At the time, 1.5°C was already considered highly optimistic. Even if 100% of all greenhouse emissions stopped by mid-century, there was already between 0.25°C to 0.5°C more warming in the atmospheric pipeline. By May of 2020, the global warming had already increased to approximately 2.16°F (1.2°C) since the industrial revolution. More than half of that increase had taken place after 1970. The rate of emissions was increasing, while environmental resilience was decreasing.

It turned out that we were also discounting the past. From the beginning of civilization, the mean global temperature had not fluctuated more than plus or minus 1.8°F (1.0°C). Prior to the great industrial acceleration that began after World War II, the ratio of cold to warm extremes was roughly one to one. By the year 2010 there were twice as many heat extremes as cold. By adding only 2.16°F (1.2°C), humanity had more than doubled the extreme warm side of the bell curve. Despite this easily measurable fact, many nations continued to discount the past by choosing not to take meaningful action to reduce emissions. Instead, they focused on adaptation and resilience. These misguided steps were a more politically convenient fit for a consumer capitalist economic policy. The problem, of course, left emissions and global warming to continue until it would overwhelm any human measures to contain it.

Some of that inaction must be accredited to Donald Trump when he withdrew America from the Paris Accord. Trump repeatedly announced that he believed that climate change was a 'hoax', despite dozens of federal agency reports. Instead of working to mitigate climate change, the Trump administration aggressively supported expanding the fossil fuel economy. He encouraged domestic use. He promoted the increased sale of coal, oil, and gas around the world. After taking office, Trump immediately removed restrictions on coal mining discharges to waters of the US. He continued his assault on environmental oversight by emasculating the US EPA. He appointed Scott Pruitt, a prima donna yes-man as EPA administrator. Mr. Trump then dismissed the EPA science advisory board.

Scott Pruitt didn't last long. He was drummed out of office for alleged intimidation of staff, waste, fraud, and abuses of authority. Trump then appointed Andrew Wheeler, another yes-man, as administrator. By March of 2020, Mr. Trump finalized cutting back proposed fuel standards for new vehicles from 55 mpg to 37 mpg. By May of 2020, Mr. Trump and his appointees were removing oversight and regulatory controls for the Clean Water Act, the Clean Air Act, and most emission standards for coal, oil, and methane gas.

In July of 2020, Trump signed an executive order to remove the requirement for an Environmental Impact Study (EIS) prior to major development projects such as the Keystone Pipeline. Fortunately, most of those cutbacks were held up in the courts until the 2020 Presidential election was over and Mr. Trump was decisively voted out of office.

Most nations continued to ignore the science and simply voted to move the goal posts. Trump and the fossil fuel lobbies were growing more successful in swaying the public away from the science. Emission reductions no longer looked at CO^2 concentrations prior to the Industrial revolution of the nineteenth century. International conferences started talking about cutting emissions to 50% of what they were in 1990. Moving the goal posts again, they proposed cutting back 50% of what they were in 1995. More recently, lobbyists proposed to cut to 50% of 2000 emissions. Those tactics only slowed action to address the climate crisis and did nothing to solve global warming. Meanwhile the pace of climate change and approaching tipping points continued their haunting acceleration.

Let me put this in perspective. If humanity was able to achieve the 2°C target, we would still be tripling the mean temperature that allowed civilization to develop during the Holocene epoch. The storms, droughts, floods, and heat extremes we experience today will be orders of magnitude worse at 2°C and unimaginably worse beyond that.

On August 9th, 2021, the 6th IPCC report, *The Physical Science Basis for Climate Change* was released. It emphatically stated that the entire global emissions output would have to reach net zero before 2050 for the world to have any chance of remaining under the 2°C target.

It is important to note that by 2022, NOAA *State Climate Summaries* reported that many of the most populated regions of the US had already reached 2°C (3,6°F) of warming. It recently became clear that the world must reach zero emissions before 2040 for the world to have any chance of remaining below 5.43°F (3°C). That projection explains why the US must be an active leader in reducing global emissions. What would happen if the US simply drifted along without leading the way to escape humanity's predicament?

Imagine you're floating down a wild river. As you enter the rapids, the riverbanks quickly steepen to become sheer cliffs. The current is picking up speed. The rapids are beginning to demand all your attention. Soon the canyon walls will be too steep to go ashore and climb out of danger. Ahead you hear the roar of rapids and see a rising mist. That roar and mist indicates a huge waterfall and certain destruction is just ahead. Slowing down will only delay disaster because the current will relentlessly take you to ultimate catastrophe. There is only one choice, paddle to shore while you can still escape before the rapids and cliffs become unsurmountable. America is complacently drifting with the current and not urgently paddling to shore.

Our own *National Climate Change Assessment* warned us that net zero by 2050 won't be enough. Global emissions are still increasing. Political lies and Big Oil special interests are still putting up roadblocks to save millions and perhaps billions of lives. At COP25, 26 and 27 there were more fossil fuel representatives than official delegates or scientists. COP28 will be held in

Dubai and chaired by an oil executive. Climate scientists and activists were moved off the main podium. Moving the goal posts only made the actions necessary to save lives more draconian and vastly costlier.

The IPCC recommendation to reduce greenhouse gases 40% by 2030 and reach net zero emissions by 2050, would still leave enough remaining GHGs in the atmosphere to continue warming the planet for centuries. Reaching net zero by 2050 would leave the US with the probability of warming at least 4.86°F (2.7°C) to 5.4°F (3°C) well before the end of this century. Even these estimates are conservative. Most model projections fail to include all the novel GHGs that continue to increase unobserved and unregulated. In October 2023, a prestigious team of scientists released a report that the present BAU scenario is inexorably leading the world toward four or five celsius degrees of warming.

The world continues drifting down the river of no return. The walls are closing in. Soon the riverbanks will be too steep to escape. We have already waited too long for moderate measures. The precautionary principle compels us to stop *all* greenhouse emissions with the urgency of a Manhattan Project well before 2050. Greenhouse gases like nitrous oxide, methane, water vapor, and other gases continue to increase at ever faster rates. We still carelessly permit fugitive emissions from fracking and methane production. Methane (CH_4) now comprises over 30% of warming. In the next decade, methane is likely to reach 35% to 50% as melting permafrost and unregulated natural gas mining emissions continue to increase.

There is a growing family of new GHGs that could surpass CO_2 or methane in the next few decades. These novel gases are man-made and do not appear in nature. Most of these novel gases are fluorocarbons and the super-hydrofluorocarbon gases used in refrigerants. One chlorofluorocarbon gas in common use is R410a. One kilogram of R410a is equivalent to two tons of carbon dioxide. More warming translates to more air conditioners, that require more refrigerant gases, that lead to more warming and another disastrous feedback loop. Thousands of other novel gases are used in manufacturing and the chemical industry. At present, there is little oversight of their use or impact on the environment, thanks to Trump's evisceration of EPA and other regulatory agencies.

The Biden administration isn't off the hook either. President Biden continues to grant new fossil fuel exploration and mining permits. He continues to subsidize the fossil fuel industry. Isn't someone telling him there is only one atmosphere.

In addition to the cessation of emissions, we will need to drawdown greenhouse gases already in the atmosphere with technologies that have yet to be invented or are nowhere near scale. We learned that delaying action with COVID was the wrong choice. Delaying action on global warming and climate change has the same effect.

A temporary small reduction in the rate of emissions began in the spring of 2020, but that was not because of good judgment or listening to science. It was the shock to the global economy caused by the COVID pandemic. There was something good learned by that rate reduction. Reducing the rate of emissions did have a noticeable impact on the atmosphere. It proved that slowing global warming can be done.

One of the stark projections by the IPCC, the US *Fourth National Climate Assessment*, and the World Health Organization (2018) was that a warming planet will cause increased risks from pandemic diseases with corresponding economic and social costs. The Trump administration's decision to withdraw funding for the World Health Organization in the middle of a pandemic illustrated his malevolence and disconnection from economic, scientific, and public health realities.

There are three sectors of society that are doing the lion's share of the work on climate change. The business community, social action groups, and young people are slowly bending the arc of humanity. Much of the business community has finally seen the cost verses benefit of decarbonizing the economy. Sustainable entrepreneurialism is late but finally off the blocks and sprinting ahead. The last two sectors include the most vulnerable to climate change, women, and children. The women's movement has joined ranks with Black Lives Matter, the Extinction Rebellion and Fridays for the Future. Governments and politicians have continued to lag far behind.

Nations can no longer ignore millions of children, teens, and young adults. Young people have learned about environmental threats in school. They may know more about the science behind climate change and environmental disruption than their parents. They understand the world wide web and how to use it. The courage, energy and tenacity of women and children cast a long shadow of shame on recalcitrant leaders. Dozens of youth leaders have spoken eloquently and often to encourage leaders to pay attention to the most current, peer-reviewed scientific knowledge.

In 2006, Economist Nicholas Stern, wrote the more than 700 page '*Stern Report—science and the impacts of climate change*'. Dr. Stern was the Chair of the Grantham Research Institute on Climate Change and the Environment, and the Economic and Social Research Center for Climate Change Economics (ESRC), and Policy at the London School of Economics and Political Science. That report was the seminal economic analysis of the global impact of climate change. His research leading up to publication indicated that it would only take 1% of global GDP to reduce emissions and keep global warming below 3.6°F (2°C).

In August of 2021 Dr. Stern commented on the significance of the August 2021 *Sixth IPCC Climate Change Report*: "I encourage all finance ministers and their ministries around the world to read this report and recognize the enormous and growing threat that climate change poses to economic development and growth. Man-made climate change will only stop once the world reaches net zero annual emissions of carbon dioxide and other greenhouse gases. Every government should now be focused on the investments and innovation required to reach net zero emissions as quickly as possible. We know that these investments will drive sustainable improvements in growth, prosperity and living standards around the world. Richer countries have a duty to help poorer countries with the finance to make the investments which will help us all."

New 2023 estimates indicate it will cost between 5.5% and 7.5% of global GDP to reach net zero emission by 2040. That might keep global warming below 4.86°F (2.7°C). If we remain on a BAU scenario until 2030, the cost of cutting emissions to net zero by 2040, will be so prohibitive it is likely to exceed the total annual operating cost of civilization. BAU beyond 2040 would be the almost certain extinction of civilization. Humanity is at the same critical juncture where all preceding civilizations have failed. We have seven years to lower global emissions by 50%. No one said virtue came easy.

NOAA has compiled state by state climate summaries for 2021. Half a dozen states had already warmed between 2°F and 3.6°F since the beginning of the twentieth century. The path forward is clear. There is no time to waste. We can start by going after the low hanging fruit of emitters. The top four sectors that produce the most greenhouse emissions are *energy, agriculture, transportation,* and *construction* in that order. There are only 100 businesses that produce 71% of all emissions. Many of these receive subsidies or special tax breaks.

There are already state, national, and multi-national plans in place. The US grid is 50 to 100 years overdue for a major overhaul anyway. Plans are outlined in numerous academic, NGO and federal study reports. Major NGO action in this sector is already underway. Why not offer a carrot to make it a coordinated effort? The stick will only be required for the most recalcitrant states and industries. Fees (taxes) on oil and gas production will be necessary. Reassignment of fossil fuel subsidies will fund new technologies. Government funded education programs will show the public that sustainable sources are already the better economic choice. North Carolina, California, Washington, Massachusetts, and Vermont are already leading the charge on renewable energy.

Sustainable energy storage has many options that do not require rare earth minerals. Battery technology is already promising grid sized capacity. Nuclear power is not dead and remains viable but slow to implement. There are safer more efficient fission technologies though fusion energy should be the eventual goal.

Fusion research has made several breakthroughs. There are three approaches to fusion presently under development. China's trillion-dollar tokamak fusion technology program may be the first to bring fusion to scale and lead the energy future of civilization. Other fusion technologies show promise with sufficient research funding by private and government sectors.

The University of Washington is working with Zap Energy on another approach. They use what is called 'sheared flow Z-pinch'. This uses powerful magnets to compress and pulse plasma to capture electricity directly. This would be more efficient than the conventional transfer of heat to turn turbines. Once thought to perpetually be thirty years in the future, fusion is now on the verge of reality.

Despite misleading promotional advertising, Big Oil and gas have stubbornly refused to diversify. It follows that they must face the consequences of greater regulation to prevent them from doing more irreversible harm. There will be screaming and gnashing of teeth, but fossil fuel companies have known about climate change and their contribution to it for more than half a century. Enough is enough. They were too big to fail—now they are too big to be allowed to continue. Corporation charters are something a nation can revoke when it threatens its security.

A huge advantage of sustainable energy is price stability for the consumer. The instability of oil and gas prices will no longer threaten household and business budgets. The transportation sector has already informed Congress that it is going electric. The government should get out of their way and encourage a full transition before 2030. Governments should initiate new public incentives to buy electric vehicles. The US government should go 100% electric by 2030. America's highway system should ensure there are enough charging stations to eliminate public concern about their EV getting stranded.

The cost vs benefits to the consumer overwhelmingly favor electric vehicles. Electric vehicles are smoother, more powerful, faster, and safer than internal combustion vehicles. Electric motors require less maintenance, last far longer, and have a flat torque curve. A flat torque curve means they generate as much power at one RPM as they do at ten thousand RPM. That makes them ideal for heavy duty or delicate work. Regenerative breaking virtually eliminates the need to replace brake pads and rotors for up to 150,000 miles under normal use. Electric vehicles do not require regular oil and filter replacement. Battery and charging issues will be a thing of the past by 2030. The benefits of electric vehicles will become far greater once national grids have made the transition to sustainable energy sources.

The cost vs benefit analysis of decarbonization has been thoroughly documented by government and NGOs for decades. Sector by sector the justification for transitioning is clear (*Drawdown,* 2017)

There should be a renewed emphasis to get mass transit and the railroad back on track with tax breaks and subsidies in strategic areas. High speed rail and several versions of underground transportation hold promise. Europe and China emphasize rail for intercity transportation and small EV trucks for intracity movement of goods.

Give air transportation a fresh look.

The construction sector must meet sustainable energy and materials specifications. Some carbon emissions can be transformed into cement and other construction materials. A significant portion of wood materials are currently cast aside as waste. That waste can be almost entirely transformed into other useful materials. Energy efficiency and generation can be incorporated into construction permits, architectural design and new building construction. There are hundreds of examples around the world of buildings that are 100% self-sufficient. Many are net energy producers.

The agricultural sector uses more water than any other. There are more mouths to feed every day. Water is already a global crisis. Land use is the second highest greenhouse gas producer. Meat production requires the most land per calorie. It consumes the most water and land for feed. Vegetable meat substitutes are gaining market share. Many people find them indistinguishable from meat and some, like me, find some preferable.

Present farming practices consume far too many pesticides and fertilizers. There are numerous proven approaches that reduce or eliminate these problems. Genetic engineering thus far has benefited corporations more than the farmer and consumer. Some crops can be modified to be perennial rather than annual. Soils are depleted of nutrients at an astonishing rate. There are many ancient and new approaches that naturally renew soils while continuing to improve food production with even greater profits.

Give sustainability a think and see what is applicable in your region. Federal and state assistance used to be an important source of guidance. Insist the government revive those practices to their fullest potential. Useful plans and strategies are everywhere.

The solutions to the four sectors of energy, agriculture, transportation, and construction are tactical. There are three overriding issues that require a more strategic approach. The elephant in the room is population. Overpopulation also has the most clear-cut solution. Make women fully franchised citizens with autonomous reproductive rights. The next issue is finding the money to accomplish global decarbonization. Money is a social construct, —it has no intrinsic value. It is only worth what we say it is worth. Its only use is in the exchange of goods and services. We already established that one hundred percent of all goods and services come from natural capital. It's the goods and services originating from natural capital that matter. For example, how much is the last drop of drinking water worth, the last acre of arable land, the last breath of air? Natural capital is the true equity behind the value of money because without natural capital, you die.

The next strategic issue is equity. Multi-million-dollar joy rides in space only show the disparity between the outrageously rich and the wellbeing of the masses. There must be a recalibration of wealth to reduce the enormous ecological footprint of the rich. A jet flying across the country emits more CO^2

in six hours than a private vehicle emits in its lifetime. One in five people go to bed hungry while the rich have lobster flown in for a supper on the veranda. When excellent health care and a quality education are only obtainable by the wealthy, some measure of redistribution is just. Our aggressive hunter-gatherer nature compels us to take a closer look at how money has become territory, and the middle class and wage earner have become the prey of the sociopath. The wealthy must pay their fair share (Robert Reich 2022).

The last strategic problem is corruption. This has a lot to do with public perception and social values. Most will agree that there is too much corruption in government, business, and politics. It is inhibiting progress on every front. All governments are corruptible, just as all forms of government are capable of good governance. I don't have an easy solution other than to say that humanity and the world we live in deserves the right to insist on truth, honesty, and virtuous governance. True freedom comes from the people not the government. When a politician refuses to act in the public's best interests, they are ignoring the US Constitution and their oath to it. Remove them.

I've tried to explain the flow of matter and energy through the systems that form civilization's global habitat. Those systems are complex and interconnected with human enterprise. Civilization is not apart from the laws of physics and the Earth's bio-geochemical systems—it is an integral part of them. Good governance must take the social, environmental, and economic pillars of sustainability under constant and equal consideration in all policy decisions and the law. The knowing and willful corrupting of our global habitat is the ultimate capital crime. It not only kills, it removes the resources for future life. The futurecide defined in this book is the premeditated sabotage of a survivable future. It is the intentional murder of future humanity and the cause of a planetary mass extinction. It is the intentional destruction of civilization through waste, fraud, and abuse.

This is the last moment for a new renaissance and a new richer and sustainable civilization.

Retrospective

It would have been easier to write as a scientist with tables and graphs to illustrate the data. Two or three of you might have read that book. I thought it best to just take you on a journey to explore where this 21st century civilization has taken us. There were many times when we had to backtrack to pick up something, some connection we missed along the way. That's how science works. We have come so far. We can see a new beginning.

Four laws of ecology appear on the first page of this book. A smart guy named Callenbach wrote them down for us: "*All things are interconnected*," like a web not a chain. *"Everything goes somewhere."* Pollution has consequences and never disappears. "*Nothing is free.*" There is always a cost. The last law tells us not to get cocky because, *"Nature always bats last."*

What we do and can do is already more than we fully understand. Our reach exceeds our grasp of the consequences. Civilization compels us to focus on the bright objects of enterprise while ignoring where it all comes from. The resulting decadence works like the proverbial monkey trap. The trap is a box with a single hole. In the box is a sweet that tempts the monkey to reach in. Once he has the sweet, his fist is too big to pull out. The monkey wants the sweet so bad that it refuses to let go. The monkey starves to death because it is trapped by its own greed. Climate change, environmental disruption and mass extinction are an existential threat to civilization, but civilization builds its own kind of monkey trap. Every civilization in history has chosen not to let go. We cling to myths, stories and the shiny objects of our own creation and ignore the reality that surrounds us.

More than sixty years ago I wondered why civilizations failed. There didn't seem to be a rational answer, and that was the answer. *Homo sapiens* is not a rational species. Civilizations are filled with greed, myth, and hubris. Civilizations build economies that favor self-indulgence and sociopathic behavior. Leaders and the general population become so sheltered from the consequences of their own decadence that they begin to choose fantasy over reality, personal gain over equity, exceptionalism over community, and power over justice. Science and technology have given the *Homo sapiens* species the power of gods. We have the power to create or destroy worlds, including this one. As a species we are only Cro-Magnon in pinstriped suits. It is that realization where the battle to save this civilization must begin.

Bibliography

Stern, N. (2006) *Stern Review on the Economics of Climate Change*, Commissioned by Government of the United Kingdom.

Dalym G.E. (1991) *Steady-State Economics*. Island Press, Washington, D.C.

Mill, J.S. (1859) *On Liberty*. Ticknor and Fields, Boston

Stiglitz, Joseph E. (2013) *The Price of Inequality*. W. W. Norton and Company Inc.

Brannen, Peter, (2017) *The Ends of the World—Volcanic Apocalypses, Lethal Oceans, and our Quest to Understand Earth's Past Mass Extinctions*. Harper Collins

Nicholas F. Gier (2016) *The Origins of Religious Violence: An Asian Perspective.* Lexington Books

Morrison, R. (1999) *The Spirit in the Gene: Humanity's Proud Illusion and the Laws of Nature*. Cornell University Press

Cunningham, William P. and Woodworth Saigo, Barbara, (1990) *Environmental Science—A Global Concern*, William C. Brown Publishers

Miller, G. Tyler, (1997) *Environmental Science*, Wadsworth Publishing Company

Morrison, R. (2013) *Origin of Faith*, Blog
Greenspan, Alan, and Wooldridge, Adrian, (2018) *Capitalism in America A History*. Penguin Random House LLC

Nace, T. (2003) Gangs of America: *The Rise of Corporate Power and the Disabling of Democracy*. San Francisco, Berrett-Koehler Publishers, Inc.

Nadeau, R.L. (2008) *Brother, Can You Spare Me a Planet?* Scientific American

Nadeau, R.L. (2008) *The Economist Has No Clothes*. Scientific American

Pianka, E.R. (2008) *The Human Overpopulation Crisis*. Addison-Wesley Longman

Pianka, E.R. (2012) *Spaceship Earth*, Addison-Wesley Longman

Pianka, E.R. (2015) *On Human Nature*, Addison-Wesley Longman

Pianka, E.R. (2000) *Evolutionary Ecology, 6th ed*. Addison-Wesley Longman

Solzhenitsyn, A.I. (1974) *Letter to the Soviet Leaders*. New York, Harper and Row

Brown, L.R. (2005) *Outgrowing the Earth*, Norton

Heinberg, R. (2003) *The Party's Over. Oil, War and the Fate of Industrial Societies*. New Society Publishers

Dowthwaite, R. (1999) *The Growth Illusion*. New Society Publishers

Catton, W.R. (1982) *Overshoot, the Ecological Basis of Revolutionary Change*, University of Illinois Press

Wells, S. (2010) *Pandora's Seed, The Unforeseen Cost of Civilization*. Random House

Flannery, T. (2002) *The Future Eaters*. Grove Press

Cohen, J. (1995) *How many People Can Earth Support?* W.W. Norton

Meadows, D., J. Randers, and D. Meadows. (2004) *Limits to Growth. The 30-year Update*. Chelsea Green Publishing Company

Ruddiman, W.F. (2010) *Plows, Plagues, and Petroleum*. Princeton University Press

Oreskes, Naomi, (2011) *Merchants of Doubt: How a handful of scientists obscured the Truth on Issues from Tobacco Smoke to Climate Change*. Bloomsbury Publishing

Oreskes, N. and E. M. Conway, (2014) *The Collapse of Western Civilization. A view from the future*. University of Columbia Press.

Pianka, E.R. and L.J. Vitt. (2019) *Our One and Only Spaceship*: Denial, Delusion and the Population Crisis.

Graedel, Thomas E. and Paul J. Crutzen, (1995) *Atmosphere, Climate, and Change*. Scientific American Library, a division of HPHLP, New York

Leopold, A. (1949) *A Sand County Almanac*. Oxford University Press

Ophuls, William, (2012) *Immoderate Greatness—Why Civilizations Fail*, CreateSpace Independent Publishing Platform, North Charleston, South Carolina

Gore, Albert, (2009) *Our Choice—A Plan to Solve the Climate Crisis*, Rodale Inc, Emmaus, PA

Gore, Albert, (2013) *The Future*. Random House

Gore, Albert, (2004) *The Assault on Reason*. The Penguin Press
Ehrlich, Paul R. and Anne H. Ehrlich, (2004) *One with Nineveh—Politics, Consumption, and the Human Future*, Island Press

Ehrlich, Paul R. and Anne H. Ehrlich, (1970) *The Population Bomb*. Sierra Club Ballantine Books

Ehrlich, Paul R. and Anne H. Ehrlich, (1990) *The Population Explosion. Simon and Schuster*

Ehrlich, Paul R., (2002) *Human Natures—genes, cultures, and the human prospect*. Penguin Books

Ornstein, Robert and Paul Ehrlich, (1989) *New World New Mind—moving toward conscious evolution*. Doubleday

Parenti, Christian, (2011) *Tropic of Chaos—Climate Change and the new geography of violence*, Norton Books

Brown, Lester R. (2012) *Full Planet, Empty Plates—The New Geopolitics of Food Scarcity*, W.W. Norton & Company

Buchan, David, (2010) *The Rough Guide to the Energy Crisis*, Rough Guides Ltd., London

Lustig, Robert H, (2017) *The Hacking of the American Mind*. Penguin Randomhouse LLC

Hoggan, James with Richard Littlemore, (2009) *Climate Cover-up—The Crusade to Deny Global Warming*. Greystone Books

Chivers, Danny, (2010) *The No-Nonsense Guide to Climate Change*: The Science, The Solutions, The Way Forward. New Internationalists TM Publications Ltd.

Smith, Laurence C. (2011) *The World in 2050—Four forces Shaping Civilization's Northern Future*. Penguin Books Ltd.
Piketty, Thomas, (2014) *CAPITAL in the twenty-first century (Summary)*. Brief and to the Point Publishing

Piketty, Thomas, (2021) *Time for Socialism—dispatches from a world on fire, 2016– (2021)* Copyright Thomas Piketty and Le MondeEditions du Seuil

Martenson, Chris, (2011) *The Crash Course*. John Wiley & Sons, Inc.

Juniper, Tony, (2018) *How We're F***ing Up Our Planet*. DK Publishing

Montgomery, David R. (2012) *The Rocks Don't Lie*. W.W. Norton & Co. Inc.

Montgomery, David R. (2007) *Dirt—The Erosion of Civilizations*. University of California Press

Montgomery, David R. (2017) *Growing A Revolution—Bringing Out Soil Back to Life*. W.W. Norton & Co. Inc.

Montgomery, David R. and Anne Bikle', (2016) *The Hidden Half of Nature—The microbial roots of life and health*. W. W. Norton & Co. Inc.

Montgomery, David R. and Anne Bikle' (2022) *What Your Food Ate—How to heal our land and reclaim our health*. W. W. Norton & Company, N.Y.

Gates, Bill, 2021, (2021) *How to Avoid a Climate Disaster*. Alfred A. Knopf a division of Penguin Random House LLC.

Carter, Peter D. and Elizabeth Woodworth, (2018) *Unprecedented Crime*. Clarity Press, Inc.

Mann, Michael E. and Tom Toles, (2016) *The Madhouse Effect*. Columbia University Press

Mann, Michael E., (2021) *The New Climate War—The fight to take back our planet*. Public Affairs Hatchette Book Group, NY

Mann, Thomas E. and Norman J. Ormstein, (2012) *It's Even Worse Than It Looks—How the American Constitutional system collided with the New Politics of Extremism*. Basic Books

McMichael, Anthony J. with Alistair Woodward & Cameron Muir, *Climate Change and the Health of Nations—Famines, Fevers, and the Fate of Populations*. Oxford University Press

Burke III, Edmund and Kenneth Pomeranz, (2009) *The Environment and World History*. University of California Press Ltd.

Sharma, Ruchir, (2016) *The Rise and Fall of Nations—Forces of change in the post-crisis world*. W.W. Norton & Co. Inc.

Meadows, Donella and Jorgen Randers and Dennis Meadows, (2004) *Limits to Growth—The 30-year update*. Chelsea Green Publishing Co.

Orlov, Dmitry, (2013) *The Five Stages of Collapse—Survivors' Toolkit*. New Society Publishers

Suzuki, David and Ian Hanington, (2012) *Everything Under the Sun—Toward a brighter future on a small blue planet*. Greystone Books

Alden, Edward, (2017) *Failure to Adjust—How Americans got left behind in the global economy*. Roman and Littlefield Publishing Group Inc.

Ellis, Richard, (2003) *The Empty Ocean*. Island Press

Ruddiman, William F., (2005) *Plows, Plagues & Petroleum*. Princeton University Press

Dawson, Ashley, (2017) *Extreme Cities—The peril and promise of urban life in the age of climate change*. Verso

Smil, Vaclav, (2017) *Energy and Civilization A History*. The MIT Press
Shiva, Vandana, (2005) *Earth Democracy—Justice, Sustainability and Peace*. North Atlantic Books

Shiva, Vandana, (2002) *Water Wars—Privatization, Pollution, and Profit*. North Atlantic Books

Shiva, Vandana, (2016) *Who Really Feeds the World?* North Atlantic Books

Shiva, Vandana, (2008) *Soil not Oil—Environmental Justice in an age of climate crisis*. North Atlantic Books

Shiva, Vandana, (1999) *Staying Alive—Women, Ecology and Development*. North Atlantic Books

Shiva, Vandana, with Kartikey Shiva, (2020) Oneness vs the 1%. Chelsea Green Publishing

Shiva, Vandana, (2020) *Reclaiming the Commons—Biodiversity, Indigenous Knowledge, and the Rights of Mother Earth*. Synergetic press

Shiva, Vandana, (2022) *Agroecology & Regenerative Agriculture—Sustainable Solutions for Hunger, Poverty, and Climate Change. Synergetic press*

Prud'homme, Alex, (2011) *The Ripple Effect—The fate of freshwater in the twenty-first century*. Scribner

Grey, Peter, (1991) *Psychology*. Worth Publishers

Sachs, Jeffrey D. (2015) *The Age of Sustainable Development*. Columbia University Press

Hawken, Paul, (2010) *Sustainable World Sourcebook—Critical Issues * Viable Solutions * Resources for Action*. New Society Publishers

Hawken, Paul, (2017) *Drawdown—The most comprehensive Plan Ever Proposed to Reverse Global Warming*. Penguin Random House LLC.

Cheon, Anerew, and Urpelainen, Johannes, (2018) *Activism and the Fossil Fuel Industry*, published by Routledge

Kolbert, Elizabeth, (2006) *Field Notes from a Catastrophe—Man, Nature, and Climate Change*. Bloomsbury USA

Klare, Michael T., (2019) *All Hell Breaking Loose*. Metropolitan Books

Quinn, Daniel, (2017) *ISHMAEL—An adventure of the mind and spirit.* Bantam Books Trade Paperback Edition

Hacker, Jacob S. and Paul Pierson, (2010) *Winner-Take-All Politics.* Simon & Schuster. Inc.

Hawken, Paul and Amory Lovins and L. Hunter Lovins, (1999) *Natural Capitalism.* Little Brown and Company

Ardrey, Robert, (1961) *African Genesis—a personal investigation into the animal origins and nature of man.* First published in the US by Atheneum

Ardrey, Robert, (1966) *The Territorial Imperative—a personal investigation into the animal origins of property and nations.* First published in the US by Atheneum

Ardrey, Robert, (1970) *The Social Contract—a personal inquiry into the evolutionary sources of order and disorder.* First published in the US by Atheneum

Ardrey, Robert, (1976) *The Hunting Hypothesis—A personal conclusion concerning the evolutionary nature of man.* New York Antheneum

Khanna, Parag, (2016) *Connectography—Mapping the Future of Global Civilization.* Random House
Alley, Richard B., (2011) *Earth—The operators' manual.* W. W. Norton & Co.

Klein, Naomi, (2019) *On (Fire)—The (Burning) Case for a Green New* Deal. Simon & Schuster

Klein, Naomi, (2017) *NO Is Not Enough*, Haymarket Books

Rifkin, Jeremy, (2019) *The Green New Deal.* St. Martin's Press

US Department of Agriculture, (1941) *Climate and Man. House Document 27, 77th Congress, 1941 Yearbook of Agriculture*

CSSR Writing Team, (2018) *Climate Science Special Report—The United States Government's Fourth National Climate Assessment (Vol. 1)*. US Global Change Research Program (USGCRP)

US Global Change Research Program, (2018) *The Climate Report—The National Climate Assessment-Impacts, Risks, and Adaptation in the United States*. Melville House

Ackerman, Diane, (2014) *The Human Age—The World Shaped by Us*. W.W. Norton & Co.

Mooney, Chris, (2005) *The Republican War on Science*. Basic Books member Perseus Books Group, NY

Nature's Operating Instructions—The True Biotechnologies, (2004) Sierra Club Books

McKibben, Bill, (2019) *Falter—Has the human game begun to play itself out?* Henry Holt & Co.

Friedman, Thomas L., (2009) (Release 2.0). *Hot, Flat, and Crowded*. Picador / Farrar, Straus and Giroux

Novacek, Michael, (2007) *Terra—Our 100-million-year-old ecosystem—and the Threats that now put it at risk*. Farrar, Staus and Giroux, New York

An Extinction Rebellion Handbook, (2019) *This Is Not a Drill*. Penguin Random House, UK

Savory, Allan and Jody Butterfield, (2016) *Holistic Management—A commonsense revolution to restore our environment (3rd Edition)*. Island Press

Rich, Nathaniel, (2019) *Losing Earth a Recent History*. Farrar, Straus and Giroux

Wallace-Wells, David, (2019) *The Uninhabitable Earth—Life After Warming.* Tim Duggan Books

Scranton, Roy. (2018) *We're Doomed. Now What?* Soho Press, Inc.

Dobzhansky, Theodosius, (1962) *Mankind Evolving—The evolution of the human species.* Yale University Press

Toler, Pamela D., (2012) *Mankind—The story of all of us.* Running Press

Wise, Timothy A., (2019) *Eating Tomorrow—Agribusiness, Family Farmers, and the battle for the future of food.* The New Press

Wolf, Naomi, (2007) *The End of America.* Chelsea Green Publishing

Eiseley, Loren, (1970) *The Invisible Pyramid.* Charles Scribner's Sons

Eiseley, Loren, (1957) *The Immense Journey.* Vintage Books a division of. Random House

Wilson, Edward O., (2012) *The Social Conquest of Earth.* Liveright Publishing Corp. Division of W.W. Norton & Co.

Wilkinson, Richard and Kate Pickett, (2009) *The Spirit Level—why greater equality makes societies stronger.* Bloomsbury Press
Stiglitz, Joseph E., (2013) *The Price of Inequality—How today's divided society endangers our future.* W. W. Norton & Company, Inc.

Rifkin, Reremy, (2011) *The Third Industrial Revolution,* Pelgrave MacMillan

Harari, Yuval Noah, (2015) *Sapiens—A brief history of humankind,* Harper Collins Publishing

Harari, Yuval Noah, (2015) *Homo Deus.* Harper

Harari, Yuval Noah, (2018) *21 Lessons for the 21st Century*. Random House Publishing Group

Diamond, Jared, (1992) *The Third Chimpanzee*. Harper Collins

Diamond, Jared, (2004) *Collapse*. Penguin Group USA Inc.

Diamond, Jared, (2005) *Guns, Germs and Steel*. W. W. Norton & Co.

Diamond, Jared, (2019) *Upheaval—Turning Points for Nations in Crisis*. Little Brown & Co.

Albright, Madeleine, (2018) *Fascism—A Warning*, Harper

Boughey, Arthur S., (1971) *Man and the Environment—An introduction to human ecology and evolution*. The MacMillan Company—Collier-MacMillan Ltd.

Masson-Delmontte, V., P. Zhai, H.-O Pörtner, J. Skea, P.R. Shukla, A. Pirani, W. Moufouma-Okia, C. Péan, R. Pidcock, S. Conners, J.B.R. Matthews, Y. Chen, X. Zhou, M.I. Gomis, E. Lonnoy, T. Maycock, M. Tignor, and T. Waterfield (eds.) (2018) *Global Warming of 1.5 degrees Celsius. An IPCC Special Report on the impacts of global warming of 1.5 C above pre-industrial levels and related global greenhouse gas emission pathways, in the context of strengthening the global response to the threat of climate change, sustainable development, and efforts to eradicate poverty*. IPCC, In House Press.

IPCC, 2021: *Climate Change 2021—The Physical Science Basis*, Summary for Policymakers.

IPCC, 2022: *Climate Change 2022—Mitigation of Climate Change*, Working Group III

EPA, 2022: *Global Anthropogenic non-CO_2 greenhouse gas emissions: 1990–2020*, US Environmental Protection Agency, Washington DC.

Environmental Change Institute (2004), *'Methane UK'*, Environmental Change Institute: Oxford, United Kingdom. Available at: http://www.eci.ox.ac.uk/research/energy/methaneuk.php

FAO (2003) *'World Agriculture: towards 2015/2030'*, Earthscan: United Kingdom. Available at: http://www.fao.org/docrep/005/y4252e/y4252e00.htm

FAOSTAT (2006) *FAO Statistics on-line database*, available at http://faostat.fao.org/site/291/default.aspx

Metz B, Davidson O, Swart R and Pan J (eds.) (2001) Intergovernmental Panel on Climate Change (IPCC) 2001: *Climate Change 2001: 'Mitigation', Contribution of Working Group III to the Third Assessment Report of the Intergovernmental Panel on Climate Change*, Cambridge: Cambridge University Press

Mosier, A. R. and Zhu, Z. L. (2000) *'Changes in patterns of fertilizer nitrogen use in Asia and its consequences for N2O emissions from agriculture systems', Nutrient Cycling in Agroecosystems*, **vol 57**, no 1, pp107–117.

Smith, P., D. Martino, Z. Cai, *et al* (2006 in press): *'Greenhouse-gas mitigation in agriculture'*, Philosophical Transactions of the royal Society, B.

WRI (2005) *'Navigating the Numbers'*, World Resources Institute, Washington DC
WRI (2006) *Climate Analysis Indicators Tool (CAIT)* on-line database version 3.0., Washington DC: World Resources Institute, available at http://cait.wri.org

Tiger, Lionel, (2017) *Men in Groups*. Routledge

Despommier, Dickson, (2010) *The Vertical Farm*. Thomas Dunne Books an imprint of St. Martin's Press

Vince, Gala, (2022) *Nomad Century—How climate migration will reshape our world*, Flatiron Books, N.Y.

Extinction Rebellion Handbook, (2019) *This Is Not a Drill*. Penguin Books Random House UK

Massive Open Online Courses (MOOCs). There are no prerequisites for any of these courses. I've taken them all and recommend these and MOOCs in general for anyone wishing to keep up to date. They can be audited free of charge or taken for credit for an average of $49.00 per course:

Environmental Security and Sustaining Peace: Presented in association with the SDG academy, Environmental Peacebuilding.

Transforming Development—The science and Practice of Resilience Thinking: Presented in association with the SDG academy.

The Age of Sustainable Development (1,2, & 3): Presented in association with the SDG academy. Presenter Jeffery Sacks, Director of the Earth Institute at Columbia University.

Climate Change in Four Dimensions: Presented by Coursera.org., Scripps Institution of Oceanography, University of California. San Diego

Denial 101x—Consensus of Scientists: Presented by edx.org.
Our Energy Future—The impact of current energy production and use, and options for a sustainable energy future: Presented by the University of California, San Diego

SUSTAINABLE DEVELOPMENT—The Post-Capitalist Order: Presented by the SDG academy; Presenter Professor Jeffrey Sacks, Director, UN Sustainable Development Network

Global Food Security: Presented by Futurelearn.com, Lancaster University (online MOOC)

Climate Change: Presented by Coursera, University of Melbourne, Australia

Climate Change—Challenges and Solutions: Presented by Futurelearn, University of Exeter, England

Climate Change: Presented by the University of Melbourne, Presenter Rachel Webster

Clouds and Climate Change: Presented by Delft Institute of Technology, Presenter Herman Russchenberg

Sustainability in Practice: Presented by Coursera, Penn State University

Globalization—Past and Future: Presented by SDG academy: Presenter Professor Jeffrey Sacks, Director of UN Sustainable Development Solutions Network.

Introduction to Sustainability: Presented by Coursera, by the University of Illinois at Urbana, Presenter Jonathan Tomkin, PhD

Climate Change—Science and Global Impact: Presented by SDG academy, Presenter, Distinguished Professor of Atmospheric Science Michael E. Mann, Penn State University

Introduction to Human Behavior and Genetics: Presented by the University of Minnesota, Presenter Professor Matt McGue PhD, Department of Psychology

Introduction to Water & Climate: Presented by edu, Presenter Professor Hubert H.G. Savenije, Delft University of Technology

Climate Literacy—Navigating Climate Change Conversations: Presented by the University of British Columbia, Presenters Dr. Sarah Burch, Dr. Sara Harris

Global Environmental Management: Presented by Technical University of Denmark, Department of Environmental Engineering, Presenter Henrik Bregnhoj

Greening the Economy—Sustainable Cities: Presented by Lund University, The International Institute for Industrial Environmental Economics, Presenter Dr. Kes McCormick

4°C Turn Down the Heat—Why a 4°C World Must be Avoided: Presented by the Potsdam-Institute für Klimafolgenforschung E.V.; Presenter Professor Hans Joachim Schellnhubr CBE, Director, Potsdam Institute for Climate Research, Potsdam, Germany

Introduction to Sustainability: Presented through Coursera by the University of Illinois at Urbana-Champaign, Presenter Jonathan Tompkins, PhD.
Global Sustainable Energy: Past, Present and Future: Presented through Coursera by the University of Florida, Presenter Wendell A. Porter, PhD.

Climate Change Mitigation in Developing Countries: Presented through Coursera by The University of Cape Town, Presenter Professor Harold Winkler, Co-Director of MAPS Program, Director Energy Research Center, U. of Cape Town

Greening the Economy—Lessons from Scandinavia: Presented by Lund University, Presenter Dr. Kes McCormick
Ecology: Ecosystem Dynamics and Conservation: Presented by the American Museum of Natural History, Presenter Dr. Ana Luz Porzecanski, Director, Center for Biodiversity & Conservation American Museum of Natural History

Climate Change Science and Negotiations: Available on YouTube, Presenter Professor Jeffrey Sacks

Sustainable Development: The Post-Capitalist Order: SDG Academy, Presenter Professor Jeffrey Sachs, Columbia University, Director UN Sustainable Development Solutions Network. How to Achieve Sustainable

Development: SDG Academy, Presenter Professor Jeffrey Sachs, Columbia University, Director UN Sustainable Development Solutions Network.

One Planet, One Ocean: SDG Academy, SDG Academy, University of Kiel- Kiel Marine Sciences

Introduction to Smart Grid: edx.org/courses/course-v1:IEEEx+2016, Modern Grid Solutions.

Planetary Boundaries and Human Opportunities: SDG Academy, Stockholm Resilience Center, Stockholm University, Presenters Professor Johan Roxström and others.

Feeding a Hungry Planet: Agriculture, Nutrition and Sustainability: SDG Academy

Cities and the Challenge of Sustainable Development: SDG Academy

Nitrogen: A Global Challenge: edX Courses, Presented by the University of Edinburgh

Link to '*The Blue Marble Report*': https://www.dougsbmr.net

Ways to reduce your carbon footprint checklist, provided by *sustainablecorvallis.org*:

DONE	NOT	CATEGORY
		Household
		Install a programmable thermostat and set the sleep and away times
		Install a heat pump water heater
		Turn off/down water heater when on vacation
		Had an Energy Trust home energy audit and implemented the recommendations
		Replace nearly all incandescent light bulbs in the house with LED or CFL bulbs
		Install insulation below the floor of your home
		Install extra insulation in the attic of your home
		Install double-paned windows
		Weather-strip windows, doors and outlets on outside walls
		Install low-flush toilets
		Install solar panels on the roof
		Use natural ventilation (open windows) for cooling instead of an air conditioner
		Unplug electronics chargers when not in use
		Hang-dry clothes on an inside rack
		Install an outdoors line for clothes drying
		Not water your lawn over the summer
		Capture roof runoff for irrigation
		Use native and drought resistant plants in landscaping
		Subscribe to Blue Sky Renewable Energy from Pacific Power
		Subscribe to Smart Energy from Northwest Natural (gas)
		Calculate your carbon footprint? https://www3.epa.gov/carbon-footprint-calculator/
		Food
		Reduce beef consumption
		Reduce meat (all kinds) consumption
		Eat vegetarian at least three nights a week
		Eat a vegan diet
		Grow a significant amount of your own food
		Strive to reduce food waste by not over-buying food and using leftovers promptly

		Carry a water bottle and not buy bottled water
		Bring your own coffee cup for take-out rather than a paper cup
		Buy local food whenever possible
		Buy organic food regularly
		Use a compost bin
		Shop at a Farmer's Market
		Belong to a community supported agriculture (CSA) subscription plan
		Belong to a cooperative grocery
	Consumer	
		Buy used goods as a first choice
		Think seriously if an item is really necessary before purchase
		Try to repair an item before disposing of it
		Choose to give gifts of time or self-effort instead of purchased
		Recycle paper, plastic, metal, and glass
		Buy food in bulk to avoid unnecessary packaging
		Participate in alternative economies such as Hour Traders, swaps, or online sites/lists
		Shop at Goodwill, Thrift Shop, The Habitat ReStore, or similar stores
		Use Energy Star ratings when selecting appliances
		Move your money from banks financing fossil fuel infrastructure—Wells Fargo, Bank of America, Chase
		Divest any investments from fossil fuel companies
	Transportation	
		Drive a hybrid car or electric vehicle
		Walk short distances rather than drive
		Ride a bike for a significant amount of in-town travel
		Use the city bus for getting downtown and back home
		Walk to work regularly
		Carpool to work regularly
		Take the city bus to work regularly
		Ride your bike to work regularly
		Telecommute sometimes
		Access Drive Less/Connect online to share car rides
		Plan that the next vehicle you buy will be hybrid or electric
		Choose to travel by train instead of a car or airplane for a trip
	Political Action	

		Show up at a demonstration supporting a renewable energy/fossil-free future
		Purchase carbon offsets? www.terrapass.com, www.nativeenergy.com, www.b-e-f.org
		Talk with friends and family about climate change issues
		Write a letter to the editor on climate change issues
		Write a letter or email to a legislator on climate change issues
		Submit comments to a regulatory agency on permitting for fossil fuel infrastructure
		Raise climate change concerns at a politician's town hall meeting
		Testify at a public hearing on climate change issues
		Vote for politicians who are committed to working on climate change
		Donate to an organization working on climate change/environmental issues
		Sign a petition related to climate change
		Subscribe to emails from a climate change or other environmental organization
		Participate in task force activities of the Corvallis Sustainability Coalition
		Attend a meeting about the Corvallis Climate Action Plan
		Attend meetings and participate in activities of a climate/environmental organization
		Join 350Corvallis for news of local events and climate activism opportunities
	Research	
		Get info on the web from 350.org, Citizens Climate Lobby, Our Children's Trust, Our Climate, Sierra Club, Climate Mobilization, Rainforest Action Network, Greenpeace, Climate Solutions, Post-Carbon Institute, Sightline Institute, Rocky Mountain Institute, Spark Northwest, Renew Oregon, Climate Reality Project, Spring Creek Project, others?

Made in the USA
Columbia, SC
13 December 2024